ENABLING TECHNOLOGIES

FOR SMART FACTORIES

Nicola Accialini

2022© Nicola Accialini

All Rights Reserved

Table of Contents

About the author V

Introduction VII

Chapter 1: Additive Manufacturing............................... 1

1.1 Brief History 2

1.2 Processes and materials 6

1.3 Hybrid manufacturing processes 26

1.4 Design for Additive Manufacturing 28

1.5 Benefits & Challenges 30

1.6 Applications 34

Chapter 2: Augmented Reality 43

2.1 Brief History 45

2.2 How AR works 46

2.3 Hardware & Software 49

2.4 Main AR smart glasses available on the market 50

2.5 Main Challenges 53

2.6 Main Applications 55

Chapter 3: Autonomous Robots... 59

3.1 Automated Guided Vehicles 61

3.2 Collaborative Robots (Cobots) 68

3.3 Drones 78

Chapter 4: Big Data Analytics... 83

4.1 Brief History 85

4.2 What Big Data are 86

4.3 Types of big data and main sources............................... 87

4.4 Analytics of Big Data 92

4.5 Main Benefits of Big Data Analytics 99

4.6 Big Data Analytics requirements and challenges.......... 101

Chapter 5: The Cloud...105

5.1 Brief History 107

5.2 Benefits 108

5.3 Limitations.. 111

5.4 Service Models 113

5.5 Industrial applications of Cloud Computing................. 126

Chapter 6: Cyber-security 129

6.1 Brief History 131

6.2 Top 5 most notorious cyber-attacks 132

6.3 Basic concepts 136

6.4 Defense methods 142

Chapter 7: Internet of Things (IoT) 147

7.1 Introduction to Internet of Things 149

7.2 Brief History 153

7.3 Industrial Internet of Things (IIoT) 154

7.4 Cyber-Physical Systems (CPS) 155

7.5 Establishing a communication 157

7.6 IoT Protocol & Standards 159

7.7 Applications of IoT 160

Chapter 8: Horizontal & Vertical IT Systems Integration .. 167

8.1 Introduction 169

8.2 Horizontal IT Systems Integration 173

8.3 Vertical IT Systems Integration 177

8.4 Inter-Organizational IT Systems Integration 178

Chapter 9: Simulation 181

9.1 Introduction 183

9.2 Discrete Event Simulation 185

9.3 Process Simulation 192

Chapter 10: Virtual Reality 195

10.1 Brief History 197

10.2 How VR works, hardware and software 201

10.3 Main benefits, limitations and applications of VR 208

Chapter 11: Other Technologies 211

11.1 Smart Human Machine Interface................................ 213

11.2 The Digital Twin... 220

11.3 Blockchain 225

Conclusion 239

List of Figures..241

List of Tables...243

About the author

Nicola Accialini is an Aerospace Engineer. After graduating from the University of Padua, he worked for some of the leading aerospace companies in international contexts. In his professional career, he has managed projects related to the development of new products and the implementation of new production technologies that in 2016 led him to take an interest in the world of Industry 4.0 and the Smart Factory.

Since June 2019 he has been living and working in Spain as a consultant and he supports companies in product and process innovation processes in the manufacturing sector.

More info: www.accialiniconsulting.com

Introduction

The concept of Smart Factory was born and developed in Germany starting from 2011 thanks to Industry 4.0, an initiative of the German government to relaunch the German manufacturing industry hit by the financial crisis of previous years.

Briefly, Industry 4.0 involves the use of a series of technologies, mainly digital, that aim to improve the efficiency and productivity of current production systems. Thanks to these technologies, it is possible to reduce the costs and times of product development, enable mass customization, reduce development and production costs and at the same time improve the quality of processes.

However, smart technologies alone are not enough. As highlighted in my previous book "Introduction to the Smart Factory", with their reasoning and learning ability, their tenacity and courage to innovate, human beings are the fundamental element of a Smart Factory: the homo sapiens (the man who thinks) and the homo faber (the man who does) use their skills and creativity to implement a factory system in which technology is used to perform dangerous, heavy and boring jobs. Men act as "deus ex machina" by designing, controlling and maintaining the production system when necessary. Only human beings will therefore be able to choose and properly use the most suitable technological tools.

In a speech to Confindustria in Bergamo (Italy) in 2017, Andrea Pontremoli, CEO of Dallara Automobili, emphasized the need to integrate skills and technology: if we take a hammer and chisel, a simple and narrow-cut technology, and we hand them over to Michelangelo, well, he will be able to create the masterpiece of "La Pietà". If, on the other hand, we give them to Andrea Pontremoli, the result will probably be "pitiful". Similarly, new technologies must be used with proper knowledge.

Another example to better clarify the concept: it is not smart to automate a production system if we do not first eliminate the waste inside it. In fact, we would create a system that automated the waste, thus making it even more inefficient. In this sense, a Lean strategy should be implemented first.

Human beings have the power, today more than ever, to use technology to their advantage in order to, if not eliminate, at least drastically reduce all those activities inside a factory that are dangerous, repetitive and alienating.

These advantages have a price to pay, namely the greater complexity of the production apparatus due to the integration of different enabling technologies.

What are these technologies, and how do they work?

The objective of this book is to provide the reader with an overview of the technologies enabling the Smart Factory, to understand their functioning, its implementation and the main challenges associated with it.

According to the Boston Consulting Group, the enabling technologies for Industry 4.0, hence the Smart Factory, are the following (in alphabetical order):

- **Additive Manufacturing**
- **Augmented Reality**
- **Autonomous Robotics**

- **Big Data Analytics**
- **Cloud Computing**
- **Cybersecurity**
- **Internet of Things**
- **IT systems integration**
- **Simulation**
- **Virtual Reality**

To them, it is necessary to add a series of other concepts that can actually be considered integrated into previous technologies, that is

- **Human-Machine Interface devices**
- **Digital Twin**
- **Blockchain**

This book is an extrapolation of the chapters of Part II of my previous book "Industry 4.0 User Guide", which aims to provide a complete overview of the principles, causes, impacts and requirements of Industry 4.0. For those who are interested, I strongly recommend it.

With this book, however, I wanted to provide users with a practical manual focused solely on the use of associated tools / technologies, at a more accessible price.

The information collected in this book is the result of both research and practical use on the field. For this reason, I will provide some ideas on how to apply them to solve real problems in a shopfloor.

Enjoy the reading,

Nicola Accialini

Chapter 1: Additive Manufacturing

1.1 Brief History 2

1.2 Processes and materials 6

1.3 Hybrid manufacturing processes 26

1.4 Design for Additive Manufacturing 28

1.5 Benefits & Challenges 30

1.6 Applications 34

Chapter Summary

Additive Manufacturing is any of various processes in which a material is joined or solidified under computer control to create a three-dimensional object, with material being added together (such as liquid molecules or powder grains being fused together). Other names to indicate this technology are Additive Layer Manufacturing or 3D Printing. Objects can be generated of almost any shape and are produced using digital model data.

In this chapter, a brief history of additive technologies will be provided and the different additive manufacturing processes, as well as the materials used, will be described. Moreover, I will present the main benefits, challenges and applications of 3D printing in various industrial and non-industrial contexts.

Keywords: additive manufacturing, 3D printing, Fused Deposition Modeling, Photopolymerization, Binder Jetting, Material Jetting, Powder Bed Fusion, Laser Sintering, Electron Beam Melting, Design for Additive

1.1 Brief History

- **1981**: Hideo Kodama of Nagoya Municipal Industrial Research Institute published his account of a functional rapid-prototyping system using photopolymers. A solid, printed model was built up in layers, each of which corresponded to a cross-sectional slice in the model.

- **1984**: On 16 July 1984, Alain Le Méhauté, Olivier de Witte, and Jean Claude André filed their patent for the stereolithography process. The application of the French inventors was abandoned by the French General Electric Company (now Alcatel-Alstom) and CILAS (The Laser Consortium). The claimed reason was "for lack of business perspective" Three weeks later, Chuck Hull of 3D Systems Corporation filed his own patent for a stereolithography fabrication system (SLA), in which layers are added by curing photopolymers with ultraviolet light lasers. Hull defined the process as a "system for generating three-dimensional objects by creating a cross-sectional pattern of the object to be formed," Hull's contribution was the STL (Stereolithography) file format and the digital slicing and infill strategies common to many processes today.

- **1988**: The technology used by most 3D printers to date (especially hobbyist and consumer-oriented models) is fused deposition modeling (FDM), a special application of plastic extrusion, developed in 1988 by S. Scott Crump and commercialized by his company Stratasys, which marketed its first FDM machine in 1992.

- **1993**: The term 3D printing originally referred to a powder bed process employing standard and custom inkjet print heads, developed at MIT in 1993 and commercialized by Soligen Technologies, Extrude Hone Corporation, and Z Corporation.

- **1995**: The Fraunhofer Institute developed the Selective Laser Melting process (SLM)

- **2002**: The year of Electron Beam Melting (EBM). Already in 1993, an application was filed by Arcam for a patent describing the principle of melting electrically conductive powder, layer by layer, with an electric beam, for manufacturing three-dimensional bodies. However, 2002 became a turning point in the history of Arcam. With two units installed at clients, the technology was finally ready for commercialization. The first production model, the EBM S12, was launched at EuroMold in Frankfurt at the end of 2002[1]

---

[1] http://www.arcam.com/company/about-arcam/history/

- **2005**: The RepRap project started in England in 2005 as a University of Bath initiative to develop a low-cost 3D printer that can print most of its own components, but it is now made up of hundreds of collaborators worldwide. RepRap is short for REPlicating RApid Prototyper.

- **2009**: In collaboration with NASA, Contour Crafting techniques are developed. Contour crafting is a building printing technology that uses a computer-controlled crane or gantry to build edifices rapidly and efficiently with substantially less manual labor. Potential applications of this technology include constructing lunar structures of a material that could be built of 90-percent lunar material with only ten percent of the material transported from Earth.

- **2013**: The company Organovo produced a human liver using 3D bioprinting, though it is not suitable for transplantation, and has primarily been used as a medium for drug testing. Organ printing was developed starting from 2003, when Thomas Boland of Clemson University patented the use of inkjet printing for cells. This process utilized a modified spotting system for the deposition of cells into organized 3D matrices placed on a substrate.

1.2 Processes and materials

Most of the material in this section has been collected by the previously mentioned book "3D Printing: Second Edition"[2] by Christopher Barnatt. For those who are interested in a deeper and more extensive understanding on Additive Manufacturing, this book is strongly recommended.

The 7 Additive processes are:

1. Material Extrusion
2. Photopolymerization
3. Material Jetting
4. Binder Jetting
5. Powder Bed Fusion
6. Direct Energy Deposition

---

[2] Barnatt C., 3D Printing: Second Edition, CreateSpace Independent Publishing Platform, 2014

7. Sheet Lamination

1.2.1 Material Extrusion

Material Extrusion refers to any 3D printing process that builds up objects layer-by-layers by putting a semi-liquid material from a computer-controlled nozzle. The most widely extruded materials are thermoplastics that can be temporarily melted for output through a nozzle. The material extrusion of thermoplastics was invented by company Stratasys that labelled the technology 'Fused Deposition Modelling' or 'FDM'. The term FDM has now become commonly used to refer to the extrusion of thermoplastics, and even to material extrusion technologies more generally. Other names for the process are 'Fused Filament Modelling' (FFM), 'Melted and Extruded Modelling' (MEM), 'Fused Filament Fabrication' (FFF) or the 'Fused Deposition Method'.

Figure 1 shows how material extrusion works: a spool of material (the 'filament') is slowly fed to a print head that is heated to between about 180 and 230°C. This high temperature melts the filament, which is then extruded through a fine nozzle. The molten filament is deposited directly onto a flat surface, the 3D printer's 'build platform' or 'print bed'. Here the filament cools and solidifies quickly, with the print head moving to trace out the first layer of the object being printed. Some material extrusion printers move the print head itself on both an upper-down and a right-left axis. Alternatively, others slide the print head back-and-forth on one axis, while moving the build platform on another. This process then repeats and repeats – often over a period of many hours – until a complete object has been printed.

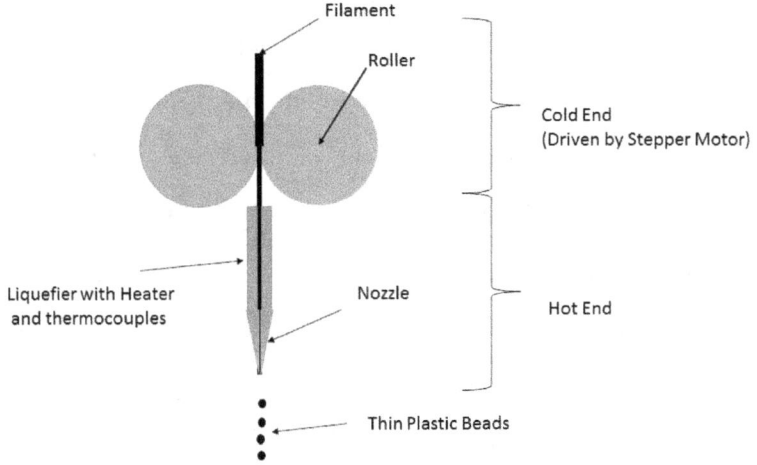

Figure 1: FDM 3D Printer Extruder[3]

Materials

ABS: Acrylonitrile Butadiene Styrene, otherwise known as 'ABS'. This is a petroleum-based thermoplastic that is widely used to injection mold a great many things like Lego bricks, cycle helmets and biros.

---

[3] https://commons.wikimedia.org/wiki/File:3D_Printer_Extruder.png

PLA: Polylactic Acid, otherwise known as 'PLA'. This is a bioplastic currently made from agricultural produce such as corn starch or sugar cane, and which is subsequently more environmentally friendly than ABS. PLA is also very safe to work with as it does not emit toxic fumes when heated.

Proto-Pasta Carbon Fiber: This is made from PLA compounded 15% by weight with very short-chopped carbon fibers. Proto-Pasta Carbon Fiber printouts are stiffer and resist bending more than standard thermoplastic parts. A similar carbon fiber filament is available from Filabot.

Laywood: Also called Laywoo-D3. Laywood is a composite of sawdust and a polymer binder, and can be melted and then extruded to 3D print objects that feel and smell like wood.

Emerging technologies used with FDM principles are:

WAAM: Researchers at Cranfield University have developed 'wire and arc additive manufacturing' (WAAM). Here a thin titanium wire is threaded through a computer-controlled, movable arm to a print head where it is heated and extruded to build up successive object layers. In December 2013 it was reported that the Cranfield team, working with their industrial partner BAE Systems, had used their WAAM technology to produce a 1.2 m spar section of an aircraft wing. This was 3D printed in titanium in just 37 hours, compared to the many weeks that would have been required for traditional manufacturing methods.

RPD: The Rapid Plasma Deposition™ (RPD™) process was developed by company Norsk Titanium. In RPD, titanium wire is precisely melted in an inert, argon gas environment. The process is monitored more than 600 times per second for quality assurance. Rapid Plasma Deposition™ technology is the ultimate in additive manufacturing. Titanium wire is melted in an inert atmosphere of argon gas and precisely and rapidly built up in layers to a near-net-shape part. The result is significantly less machining, and ultimately, a 50%–75% improvement in buy-to-fly ratio compared with conventional manufacturing methods.

1.2.2 Photopolymerization

"While almost all consumer 3D printers are currently based on material extrusion, many industrial 3D printers use more accurate if more expensive processes that bind powders,

solidify liquids, or bond sheets of material together. The first category of these technologies goes under the generic heading of 'vat photopolymerization', and uses a light source to solidify successive object layers on the surface or base of a vat of liquid photopolymer."[4]

Vat photopolymerization is already commercially achieved via five distinct methods known as

- Stereolithography (SLA)
- Digital Light Projection (DLP)
- Scan, Spin and Selectively Photocure (3SP)
- Lithography-based Ceramic Manufacturing (LCM)
- Two-photon Polymerization (2PP)

For brevity, the most common process, the SLA, is following described.

As previously stated, Stereolithography was the first ever 3D printing process developed by Hideo Kodama, and uses a computer-controlled laser beam to build a 3D object within a vat (or tank) of liquid photopolymer.

---

[4] Barnatt C., 3D Printing: Second Edition, CreateSpace Independent Publishing Platform, 2014

Figure 2: How Photopolymerization works shows how Photopolymerization works. In most SLA printer, objects are built on a perforated build platform which is initially positioned just under the surface of a photopolymer vat. A UV laser beam then traces out the shape of the first object layer on the surface of the liquid. This causes it to 'cure' (set solid), then the build platform lowers just a little. More liquid photopolymer then either naturally flows over the top of the first object layer, or is forced across it by a mechanical mechanism, and the next object layer is traced out and set solid by the laser. This process then repeats over and over until the whole object has been printed.

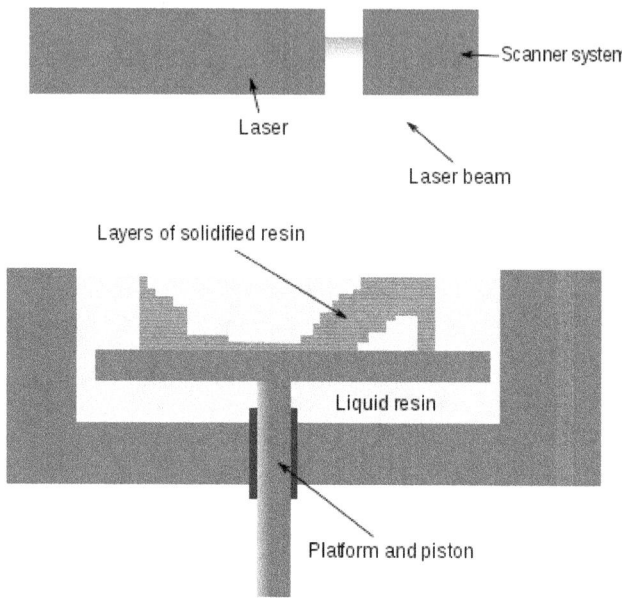

Figure 2: How Photopolymerization works[5]

Materials

Today a wide variety of materials have been developed, including rubber-like plastics, many substitutes for ABS and other thermoplastics, flame retardant plastics, clear resins,

---

5

https://commons.wikimedia.org/wiki/File:Stereolithography_apparatus_vector.svg

and special photopolymers for dental modelling and jewelry design.

Material extrusion VS Photopolymerization

A qualitative comparison between the two main technologies available to consumer is shown in the following table:

Feature	*Material Extrusion*	*Photopolymerization*
Resolution	about 0.1 mm	less than 0.05 mm
Volume	Larger	Smaller
Speed	Slower	Faster
Material cost	Lower	Higher
Material variety	High	Low

Table 1: Material Extrusion VS Photopolymerization

1.2.3 Material jetting

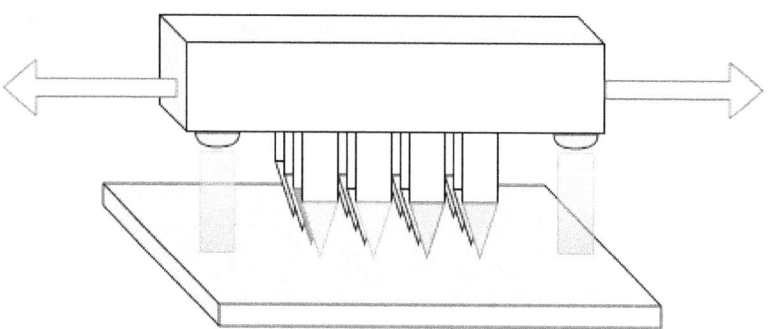

Figure 3: Schematic representation of Inkjet Technology[6]

Material Jetting is another 3D printing technology based on the solidification of liquids.

Material jetting exists in various formats, most of which spray a liquid photopolymer from a multi-nozzle print head. The print head moves across the build platform depositing one layer of liquid photopolymer which is then set solid with UV light also emitted from the print head. 3D printer manufacturer Stratasys sells hardware based on this process under their trademarked name 'PolyJet' (short for

---

[6] https://commons.wikimedia.org/wiki/File:Inkjet_3D_Printing.svg

'photopolymer jetting'), while 3D Systems have labelled the technology 'MultiJet Printing' or 'MJP'."

Materials

A wide range of materials have been created for material jetting, including compounds that simulate the properties of ABS, polypropylene, polycarbonate and rubber. Printers also have the capability to output multiple materials in the same print job by supplying different photopolymers to the print head, and by mixing them in different combinations during the printing process.

4.2.4 Binder jetting

Figure 4: How Binder Jetting works shows how Binder Jetting works. The process starts when a layer of powder is laid on a build platform called the "powder bed". This is usually achieved by raising the base of an adjacent "powder reservoir" and using a sweeper blade to push the powder across the bed. A multi-nozzle print head then travels across the powder bed, selectively jetting a binder solution onto it in the shape of the first object layer. The powder bed is then lowered, another layer of powder is laid down, another layer of binder is jetted onto it, and so on. After the process is completed, a depowdering operation must be processed to remove the powder in excess, typically by compressed air.

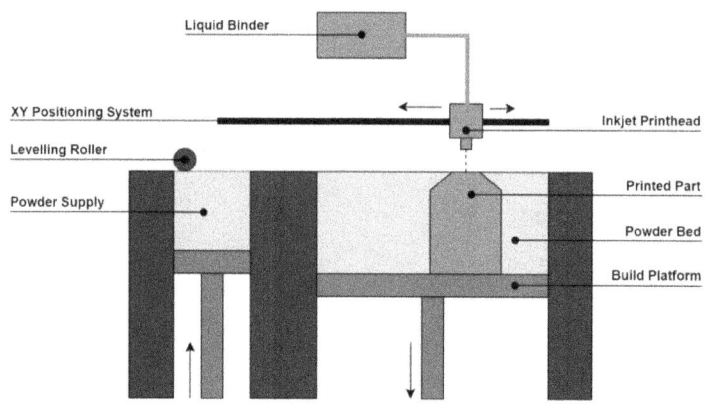

Figure 4: How Binder Jetting works[7]

---

[7] https://commons.m.wikimedia.org/wiki/File:Binder_jetting.png

Benefits of Binder jetting are:

- no support structures have to be printed or removed, as overhangs or orphan parts are always supported by the loose powder that surrounds an object while it is being printed;
- the process is typically faster than other 3D printing methods;
- they are capable of outputting objects in full colour. To make this happen, binder jetting sprays coloured inks as well as binder solution onto each layer of powder. The technology is exactly the same as that used in traditional, 2D photo printers, with cyan, magenta, yellow and black inks applied in an appropriate combination to produce full colour printouts.

On the other hand, it is not possible to make object 100% solid. To make this happen, it is necessary to turn to Powder Bed Fusion 3D printing technology, which will be described later on.

Materials

Polymer: 3D Systems sells a full-colour binder jetting printer called ProJet 4500 that makes objects from a plastic powder called VisiJet C4 Spectrum. This 3D printer was launched in December 2013 and offers the ability to create semi-rigid plastic parts that require no post-processing.

Sand: some printers can also use casting sand, which allows molds and interior mold sections (the "cores") to be 3D printed, with significant industrial applications. The main benefit of using 3D printing for sand cast molds is that there is no need to create a physical pattern of an object before it is printed (a process that requires time and craft skills). The application of this technology can therefore save a great deal of time and money, as well as allowing novel products to be created.

Metal: bronze, iron or stainless-steel infused with bronze, or nickel-based alloy parts are possible. Parts made by Inconel 625, for example, are used in aerospace to make turbine blades and other high-end industrial components.

Ceramic: the same process used for metals can be used with ceramic materials, such as alumina silica ceramic.

Glass: ExOne has binder jetting hardware to create glass objects.

2.2.5 Powder Bed Fusion

Powder Bed Fusion (PBF) methods use either a laser or electron beam to melt and fuse material powder together. In the following sections we will provide an overview of two of the most common PBF processes: Laser Sintering and Electron Beam Melting.

2.2.5.1 Laser Sintering

In Laser Sintering, also known sometimes as "Laser Beam Melting" (LBM), a layer of powder is rolled (or swept) across a powder bed, following which a laser beam traces out the cross-section of the first object layer. The heat from the laser "sinters" the powder granules that it touches, so causing them to at least partially melt and fuse with adjacent granules.

There are different trademarked variants of laser sintering to use laser beam to fully melt the granules of a single-material powder in order to produce purer metal objects. These processes have some small differences depending on their implementation and the particular manufacturer. Here a list of different laser sintering processes:

- Direct Metal Laser Sintering (DMLS)
- Direct Metal Printing (DMP)
- Selective Laser Melting (SLM)
- Micro-Laser Sintering (MLS)

The surface quality of objects produced via laser sintering is excellent, and a range of finishes can be achieved. The main limitation of laser sintering is the expensive hardware, with each printer typically costing several hundred thousand dollars.

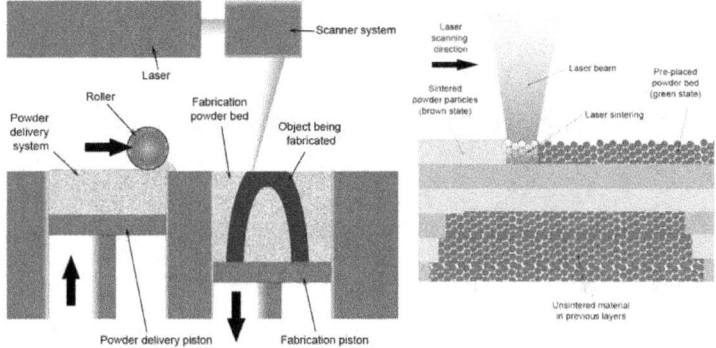

Figure 5: How Laser Sintering works[8]

Materials

A wide range of materials can be used, including plastics, many metals, ceramic, sand and wax. Laser sintering is a very accurate process and produces excellent results in plastic materials such as nylon. However, it is not possible to build objects with mechanical properties suitable for some engineering applications like engine components.

8

https://commons.wikimedia.org/wiki/File:Selective_laser_melting_system_schematic.jpg

4.2.5.2 Electron Beam Melting

Electron Beam Melting (EBM) is also known as Electron Beam Additive Manufacturing (EBAM). The process has been invented by company Arcam, and builds objects layer-by-layer in a vacuum, with electron beam making multiple passes of each object layer. The electron beam moved around via electromagnetic deflection, rather than being directed by mechanically -mechanized mirrors.

The main benefit of EBM over LS is that completely dense metal parts can be accurately created with zero distortion. EBM is used for aerospace and other specialist industrial sectors. Medical implants have also been produced using this technology. The main limitation at the moment is the limited build volume, around 350 x 350 x 380.

2.2.6 Direct Energy Deposition

In Directed Energy Deposition (DED), metal powder is directed into a high-power laser beam for deposition. Unlike in powder bed fusion, the metal powder fed to the print head can be altered continuously during printout. Directed energy deposition can therefore fabricate objects with properties that cannot be obtained using traditional production methods.

Because a flat powder bed is not required, directed energy deposition has the advantage of being able to repair existing objects.

Materials: stainless steel, copper, nickel, cobalt, aluminum and titanium.

2.2.7 Sheet lamination

Sheet Lamination was invented by a company called Helisys in 1991 that has since gone out of business. Sheet lamination refers to a variety of mechanisms, but typically all advance a sheet of build material onto a build platform. This material may have an adhesive backing or have adhesive applied during the process. A laser (or blade) is used to cut the outline of an object layer into the sheet, and the build platform lowers just a little. The process then repeats until all object layers have been created.

Materials

This technology sticks together sheets of paper, plastic or metal foil that are shaped into object layers by cutting them with a laser or blade. Where sheet lamination adheres sheets of paper, it is also known as "laminated object manufacture" (LOM).

1.3 Hybrid manufacturing processes

Hybrid processes refer to the combination of additive and subtractive manufacturing processes applied sequentially or integrated. Several major CNC manufacturers have already introduced bolt-on 3D print heads for their platforms: companies producing new 3D printer systems have likewise begun to incorporate features allowing limited subtractive machining in their processes. The arrival of commonplace "hybrid" manufacturing is probably inevitable, as improvements both in CNC and in 3D printing create larger areas of capability overlap between the two.

Hybrid manufacturing offers a number of benefits over machining or 3D printing alone because of additive's ability to add material just where it is needed. 3D printing also allows for the use of multiple materials in a single part, which makes it possible to clad weaker metals for greater strength, add copper to aid heat transfer or save money on material by applying expensive metals only where they are needed. When additive is combined with machining in a hybrid system, it is possible to 3D print and finish a part in a single setup. This approach reduces error because the printed part does not have to leave the build envelope and be reset on a separate machine. It is also possible to alternate printing and machining to finish internal features, such as conformal cooling channels for injection molds, or features that would be inaccessible in the completed part[9].

---

[9]https://www.additivemanufacturing.media/articles/am-101-hybrid-manufacturing

Another interesting application of hybrid manufacturing is in repair: for example, if a critical feature is machined below the lower dimensional limit, material can be added via 3D printing and subsequently removed via CNC tools to reach the final dimension.

1.4 Design for Additive Manufacturing

With subtracting machining and standard processes, design is mainly driven by manufacturing processes. Which means, it is not possible to create very complex/ideal geometries for different reasons:

- complex shapes can be simply not feasible
- complex shapes are often more expensive

As a relatively new manufacturing technology, engineers and designers have little experience and insufficient knowledge on the capabilities and limitations of AM. Design for Additive Manufacturing (DfAM) have been recently introduced as one of the supplementary design tools for selecting the process parameters (such as cost, time, quality, reliability and CAD constraints) in an optimal manner.

Because "complexity is free", adding complex features and geometries is feasible and doesn't add costs. Topology optimization is a method for obtaining the best possible geometry/shape while satisfying certain requirements. For example, optimization of product's volume while maintaining minimum compliance of parts have frequently investigated. For lower weight-to-stiffness ratio, there exists several novel topology optimization applications.

Another example of a new way of design thinking is Design for Assembly (DfA). In most cases, mechanisms require 2 or more parts to be assembled. Using AM properly, now it is possible to print element and mechanism without any assembly need.

1.5 Benefits & Challenges

Benefits

"I always find it difficult to look at one specific added value that additive manufacturing brings. The fact that you manufacture additively instead of subtractively per definition, in my opinion, always gives you lightweight solutions [...] Lighter weight for me goes hand-in-hand with better stiffness, more complexity, or assembly reduction, so I struggle when I am asked to narrowly focus on a specific added value with this technology. If as a designer I have a new toolbox that can increase the performance of a system five different ways, why would I only focus on one? You can absolutely lightweight parts with additive, but you can do a lot more." Koen Huybrechts, Senior Project Engineer, 3D Systems

Lightweight: nowadays, lightweight is a critical factor in new product design. In fact, lightweight components have several benefits: 1) Lightweight cars or airplanes mean less fuel consumption, which leads to less fuel cost and less pollution; 2) Less material required, which can lead to cheaper components; 3) Buy-to-fly ratio, which means less waste.

Feasibility: with AM, there are almost no design limitations, and components that before were simply impossible to produce, now can be manufactured.

Complexity is free: complex components are traditionally manufactured by casting or milling centers; however, depending on their complexity, costs, set up and processing time may vary significantly. With AM, it doesn't matter how complex a component is, there are no additional cost due to design complexity.

Cost: depending on the method and requirements, parts can be much cheaper than traditional processes. This is particularly true with FDM. For example, Product Development consists of an iterative approach that can involve different design iterations. Each design iteration may imply new tools, fixtures, supports etc. that can be potentially used only few times before the new design iteration. In these scenarios, FDM with polymeric materials may be strong enough and even 80-90% cheaper than traditional items. Another application where AM can result cheaper than traditional processes is Low Volume: indeed, developing a few numbers of items can be extremely expensive as all the development costs should be charged on them.

Rapid prototyping: AM provides a cheap and fast solution to make prototypes. Prototypes are useful in different stages of Product Development, from design concept generation to buy off final products.

Faster: making fixtures, supports or masks to support the production process may involve several weeks, from design to manufacturing and shipping. With AM, once the 3D model is ready, the printing process may take hours, maybe days, depending of the size and accuracy. In any case, AM may reduce the lead time from weeks or months to hours or days. In any case, it provides more time to the design phase the most value-added step.

Flexibility: AM can print any shape. This means that no additional tooling or fixturing are required, no additional set up or reprogramming.

Personalization: Due to its high flexibility, AM can personalize products. This is particularly useful in fashion businesses, where customers may choose their jewelry depending on their personal taste, or in manufacturing, where selective assembly may be required.

Challenges

Knowledge: The first form of AM was invented in 1981, therefore can't be considered a new technology. However, the knowledge among industries is still quite low, as its potential. AM is considered most of the time a DIY technology for people who want to replicate small plastic objects at home, or, on the other hand, AM is considered something to be used only in high-tech industries. This is definitely a mis-concept: some forms of AM are very cheap and can provide big benefits to Small Middle Enterprises (SME). For final products, although 3D printing components have been already used for high-demanding applications like Formula 1 or aerospace, it is still at quite an early readiness level for mass production applications.

Size: The build volume really depends on the technology to be used, but in general it is not wrong to say that it is quite limited.

Mechanical properties: Although some 3D printed components have been using for high-demanding applications, however in some cases mechanical properties are a limitation.

Cost: Although cost can be cheaper for some application and AM technologies, however it can be extremely high for specific technologies. This can be true when all powder-based technologies are used, as machines and powders are very expensive and the process itself time consuming.

1.6 Applications

Industrial Applications

Rapid prototyping: it is the most popular application. A Replica can be very useful in the following situations:

- show the final product to the customer;
- better evaluation of critical features;

- NC programming to evaluate real space and collisions (e.g., CAM, CMM etc.)
- show features to suppliers for a better understanding

Final products: According to Barnatt, over 20 per cent of 3D printed objects are already end-use components, but by 2020 this figure is expected to rise to at least 50 per cent[10]. In 2016, GE entered a new engine into service called CFM LEAP. This will feature a combustion component that it is not possible to produce using traditional manufacturing methods. Remaining in the aerospace sector, the Boeing 787 Dreamliner has 30 3D printed parts, including air ducts and hinges.

Tools & Fixtures: Typically, manufacturers create tooling in metal as needed by machining it in house or outsourcing it. Depending on the forces experienced by the part, it may not always be necessary to produce these tools in metal. SLA 3D printing materials have advanced significantly, and there are a number of functional resins well suited to 3D printing jigs and fixtures. 3D printed tools and fixtures can be really useful especially in product development. Indeed, plastic materials

---

[10] Barnatt C., 3D Printing: Second Edition, CreateSpace Independent Publishing Platform, 2014

are nowadays strong enough for a wide range of application, including:

- Fixturing for final inspection (e.g., visual inspection, optical inspection, roughness check, CMM)
- **GO-NOGO gauges**, limited to the tolerances required
- **Support for NDE** (e.g., Magnetic Particle Inspection, Nital Etching, Fluid Particle Inspection, Barkhausen etc.)
- **Small tools for manual operations**, like sealing, gluing and son on;
- **Machine set up**: calibration jigs as well as machine repair and maintenance tools improve line startup efficiency when getting production up to speed
- **Fixturing and positioning**: soft jaws, assembly jigs, and other work-holding devices require alignment features that can be hard to machine

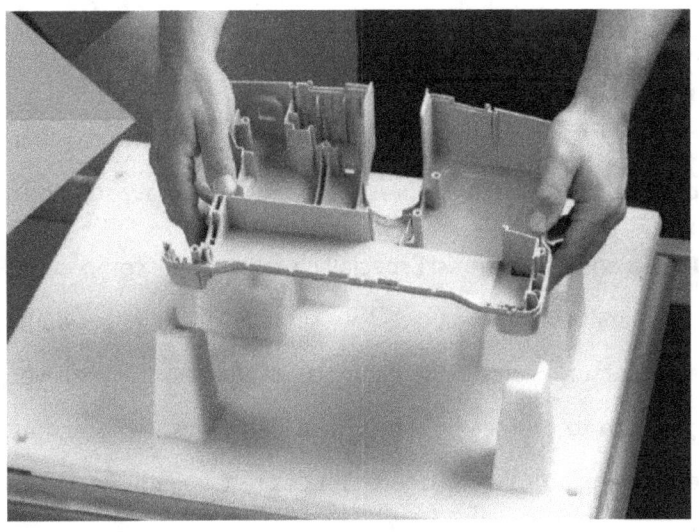

Figure 6: Example of 3D printed support for final inspection[11]

Shadowboards: They are useful to keep the workspace organize and it is part of 5S techniques. Typically produced with foam, shadowboards can be easily produced with 3D printing techniques as well, especially in conditions of high-cleanliness level requirements

---

[11] Stratasys white paper, 3D PRINTING JIGS & FIXTURES FOR THE PRODUCTION FLOOR

Maintenance: In some situations, machines are stopped because a simple connector, socket, tube is broken and can take days or sometimes even weeks before receiving a spare part. 3D printed items represent a valid temporary solution

Masks and protections: Especially for high-value components, it is important to avoid any scratch or damage during manufacturing steps. For example, in gear industry is it very common to protect critical surfaces with plastic protections. depending on volumes, these can be pretty expensive, as they are typically produced by injection molding, with the mold milled in a milling centre. 3D printed masks and protections can save a significant amount of time and reduce costs. Moreover, similar protections are required in special processes like surface finishing operations. Shot peening, for example, requires rubber mask to protect areas where the treatment is not required.

Molds: in previous section, we have seen how binder jetting can use sand to create sand casting molds. Other materials and technologies can be exploited

Other Applications

3D printed vehicles: 3D printing has already been used to help produce two working automobiles. The first of these is called "Urbee", created in 2011 by Jim Kor and his company Kor EcoLogic. Urbee is a low-energy, 2-passenger hybrid vehicle[12]. The engine and most interior components are not 3D printed, however Urbee's body shell was built layer-by-layer on Stratasys hardware in 10 thermoplastic sections and took about 2500 hours to print. Another example of 3D vehicle is a bike frame printed by the UK company Renishaw, which has been featured in the 2015 edition of Guinness World Records for manufacturing the world's first 3D printed titanium alloy bicycle frame. The project was completed in collaboration with British bike manufacturer Empire Cycles.

3D printed surgical models: at Kobe University School of Medicine in Japan, Maki Sugimoto is using 3D printers to create medically-accurate models of patient kidneys, livers

---

[12] Barnatt C., 3D Printing: Second Edition, CreateSpace Independent Publishing Platform, 2014

and other organs, derived from MRI and CT scan data, and can prove invaluable in planning surgery. Model organs are created used material jetting 3D printers that can build things in multiple materials.

3D printed prosthesis: According to a statement made by the American Orthotics and Prosthetics Association, the average prosthetic costs between $1,500 to $8,000. This expense is often paid out of pocket rather than covered by insurance. By contrast, a 3D printed prosthetic costs as little as $50! 3D printed prosthetics can also be made much quicker; a limb can be made in a day. Furthermore, consumers can easily customize their purchases, which is another enticing factor for kids. Children can pick out colors and styles to fit their wants and needs.

Bioprinting: The US company Organovo produces a 3D printer called the Novogen MMX that 3D prints layers of living human cells, and is at the forefront of research into organic "bioprinting". The Bioprinting of replacement human tissue is without doubt the most radical application of 3D printing so far conceived. Technically speaking, bioprinting is a bioadditive manufacturing technology and an extension of the science of tissue engineering.

Chapter 2: Augmented Reality

2.1 Brief History 45

2.2 How AR works 46

2.3 Hardware & Software 49

2.4 Main AR smart glasses available on the market 50

2.5 Main Challenges 53

2.6 Main Applications 55

Chapter Summary

> *"According to the Augmented / Virtual Reality Report 2017 (Digi-Capital, 2017), the VR/AR market should reach $108 billion revenues by 2021, while an overview by Goldman Sachs (2016) suggests a business of $80 billion in 2025"*[13].

In this chapter, we will see how Augmented Reality evolved through the years, how it works, what hardware is required,

---

[13] Gandolfi E, Handbook of Research on K-12 Online and Blended Learning (2nd ed.), Publisher: ETC Press, Editors: Kennedy, K, Ferdig, R.E., pp.545-561

how to develop the required algorithms and what smart glasses are available on the market: In the last part, we will present some main barriers in using this technology and what the main applications are.

Keywords: Augmented Reality, AR, Smart Glasses

2.1 Brief History

- **1968**: Ivan Sutherland invented the first VR head-mounted display at Harvard University, called "the Sword of Damocles"

- **1984**: In the movie "The Terminator", the Terminator uses a heads-up display in several parts of the film. In one part, he accesses information about his "new" motorbike

- **1990**: The term 'Augmented Reality' is attributed to Thomas P. Caudell, a former Boeing researcher

- **1994**: Julie Martin creates "Dancing in Cyberspace", the first augmented reality theater production. In "Dancing", acrobats dance in and around virtual objects on stage.

- **1999**: Hirokazu Kato created ARToolKit at HITLab

- **2013**: Google announces an open beta test of its Google Glass augmented reality glasses. The glasses reach the Internet through Bluetooth, which connects to the wireless service on a user's cellphone. The glasses respond when a user speaks, touches the frame or moves the head.

- **2015**: Microsoft announces Windows Holographic and the HoloLens augmented reality headset. The headset utilizes various sensors and a processing unit to blend high definition "holograms" with the real world

- **2017:** IKEA released its augmented reality app called IKEA Place that changed the retail industry forever

2.2 How AR works

Augmented Reality works according to one of the following approaches:

1. **SLAM:** Simultaneous Localization and Mapping (SLAM) is the computational problem of constructing or updating a map of an unknown environment while simultaneously keeping track of an agent's location within it. There are several algorithms known for solving it. Published approaches are employed in self-driving cars, unmanned aerial vehicles, autonomous underwater vehicles, planetary rovers, newer domestic robots and even inside the human body.

2. **Recognition Based:** recognition (or marker) based augmented reality uses a camera to identify visual markers or objects, such as a QR/2D code or natural feature tracking (NFT) markers, to showcase an overlay only when the marker is sensed by the device. Marker-based AR technology depends upon device camera to distinguish a marker from other real-world objects. Not only the marker image but the position and orientation can also be calculated. Once recognized, the marker on screen is replaced with a virtual 3D version of the corresponding object. This is done to permit the user to observe the object in more detail and from various angles.

Rotating the marker would rotate the virtual replication as well.[14]

3. **Location Based:** unlike recognition based, location-based AR relies on GPS, digital compass, velocity meter, or accelerometer to provide data about the location and the augmented reality visualizations are activated based on these inputs. It is also known as markerless augmented reality. The location detection features in smartphones make it easy to leverage this type of augmented reality technology, making it quite popular. Some common uses of location-based AR include mapping directions, finding nearby services, and other location-centric mobile apps. A popular example is provided by Pokémon Go. The game uses your phone's GPS, camera, and clock to generate a brightly-colored, fantastical version of the real world on your phone[15]. Much like Google Maps, "Pokémon Go"

---

[14] Know the Augmented Reality Technology: How does AR Work?
https://www.newgenapps.com/blog/augmented-reality-technology-how-ar-works

[15,54] Know the Augmented Reality Technology: How does AR Work?
https://www.newgenapps.com/blog/augmented-reality-technology-how-ar-works

tracks your phone's location using GPS, integrating this information with an in-game map.

2.3 Hardware & Software

Hardware components for augmented reality are:

- processor
- display
- sensors
- input devices

Displays are typically smart glasses, smartphones and tablets. Nowadays, tablets and smartphones are also very common and widely used with AR. These devices have all features required: portable, powerful CPU, digital camera, GPS, inertial sensors and a touch screen. According to Time Magazine, in about 15–20 years it is predicted that Augmented reality and virtual reality are going to become the primary use for computer interactions.

The software must derive real world coordinates from camera images. That process is called image registration, and uses different methods of computer vision. Augmented Reality Markup Language (ARML) is a language specifically developed for AR. To enable rapid development of augmented reality applications, some software development kits (SDKs) have been developed. Some AR SDKs are offered by:

- Vuforia
- ARToolKit
- Catchoom CraftAR Mobinett AR
- Wikitude
- Meta
- ARLab

2.4 Main AR smart glasses available on the market

Google Glass: while it failed to find purchase in the consumer market, Google's foray into Augmented Reality has seen considerably more success in medical and manufacturing applications. The device features some ambitious technology, including bone conduction for audio output and a prism projector display.

The successor to the Explorer Edition, Google Glass Enterprise Edition eschews the bone conduction tech, but also makes several improvements over previous iterations, including an Intel Atom processor and a barometer.

Microsoft HoloLens: billed as the world's first full untethered holographic computer, the HoloLens is available in both a Development Edition and a Commercial Suite configuration. Companies based in North America also have the option of renting a HoloLens through a Microsoft partner. The current version of the HoloLens has a 35° FOV, but a recent patent from Microsoft for a MEMS laser scanner could increase the FOV for the next generation of HoloLens to 70°.

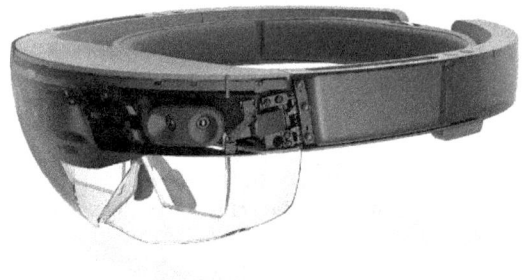

Figure 7: Microsoft HoloLens Headset[16]

VUZIX M100 & M300 Smart Glasses: founded in 1997, Vuzix has a long history in the augmented reality industry. Its current offerings suitable for industrial applications are the M100 and M300 Smart Glasses. The M300 is marketed as having enhanced functionality, and there are some important differences between the two. For example, the diagonal FOV on the M100 is 15° and 20° on the M300.

DAQRI Smart Glasses: the latest iteration of the company's augmented reality hardware, DAQRI's Smart Glasses are

---

[16] https://commons.wikimedia.org/wiki/File:Ramahololens.jpg

designed for professional use. The unit is based on a 6th generation Intel Core m7 processor and uses dual LCoS optical displays, which give it a 44° diagonal FOV. It comes equipped with Worksense Standard, DAQRI's suite of apps for AR tasks.

2.5 Main Challenges

User Interface: one of the main challenges of AR is to improve field of view (FOV), brightness, display quality, latency, etc. In the case of FOV, for example, even the best AR HUDs can only offer up to 90 degrees. Compare that to the 190-degree horizontal and 120-degree vertical FOVs for normal human vision and the gap between where the technology is now and where it needs to go becomes obvious.

Health issues: intense use of VR headset might cause some health issues like:

- **Virtual Reality Sickness:** VR headsets contain 2 small LCD monitors, each projected at one eye, creating a stereoscopic effect which gives users the illusion of an artificial environment created all around you. And longer time of screen focus increase feeling of nausea and distortion, a phenomenon called Virtual Reality Sickness

- **Eye Strain:** While using AR/VR headsets we are confined in a limited area from our environment. And we see and focus on particular details for long due to which we may tend to blink less compared to normal frequency of blinking. This results in drying of the front surface of our eyes and strain on eyes
- **Dizziness:** Individuals having amblyopia (an imbalance in visual strength between the 2 eyes) or other conditions inhibits focusing, depth perception or normal 3D vision may not experience 3D effects of VR headsets. Individuals with these disorders may be more likely to experience headaches and eye fatigue when using VR gear

Lack of knowledge: There is still quite a lack of knowledge about this technology and an insufficient level of digitalization within the company

2.6 Main Applications

Facility planning: facility planning is a high risk and high investment process. Layout design is typically performed in 2D or in a 3D CAD environment, which in most case represent a "good enough" approach. On the other hand, other issues like ergonomics can't be easily assessed without expensive simulation software (e.g., Jack by Siemens). AR can help the factory planner to experience how new machines or new production lines fit in the existing space and if workbenches and conveyors have been designed at the right height to avoid ergonomic issues.

Smart drawings: especially for components with complex geometry, AR may help to understand main features of a drawing. AR may help to understand how parts should work in services, highlighting critical features and adding more info like specifications, materials, method of manufacture and method of assembly. Possibilities are limitless.

Learning: similar to smart drawings, AR may increase the learning effectiveness by providing more realistic experiences in the learning process.

Product Development Process: At different stages of product development, AR may support the developer in decision making by creating a unique experience at lower cost. Different design solutions and modifications, for example, may be experienced on existing products.

Quality inspection: AR can support quality inspector during their normal activities. These types of solution are becoming more and more popular.

Assembly instructions: like quality inspection, AR supports workers in the factory to assemble complex components. Instructions like torque values, assembly sequences, tooling to use are displayed to guide the fitter step by step through the entire process

Maintenance Repair and Overhaul (MRO): augmented reality can enhance a worker's ability to perform maintenance routines by superimposing simple step-by-step instructions on their field of view. This is the logical next step in being able to see machine status at a glance using AR, which is a benefit that goes beyond MRO applications.

Plant Maintenance: in a similar way, instructions for maintenance procedures may be provided.

Chapter 3: Autonomous Robots

3.1 Automated Guided Vehicles 61

3.2 Collaborative Robots (Cobots) 68

3.3 Drones 78

Chapter Summary

Robots are the result of the third industrial revolution. The term comes from a Czech word, "robota", meaning "forced labor"; the word 'robot' was first used to denote a fictional humanoid in a 1920 play R.U.R. (Rossumovi Univerzální Roboti – Rossum's Universal Robots) by the Czech writer, Karel Čapek but it was Karel's brother Josef Čapek who was the word's true inventor.

Industrial robots are becoming a key technology in smart factories not only because of their increasing capability and flexibility, but also because they are becoming constantly cheaper (see Figure 8). In this chapter, 3 categories of robots are described: Automated Guided Vehicles, Collaborative Robots and Drones.

Keywords: Robots, Automated Guided Vehicles, AGV, Collaborative Robots, Cobots, Drones

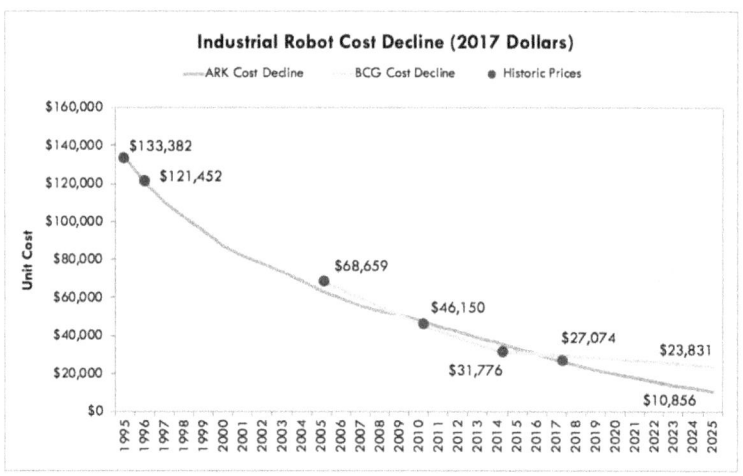

Figure 8: Industrial robot cost decline[17]

---

[17] https://ark-invest.com/articles/analyst-research/industrial-robot-cost-declines/

3.1 Automated Guided Vehicles

An Automated Guided Vehicle (AGV) is a portable robot that follows markers or wires in the floor, or uses vision, magnets, or lasers for navigation. They are most often used in industrial applications to move materials around a manufacturing facility or warehouse. Applications of the automatic guided vehicle broadened during the late 20th century.

The first AGV was brought to market in the 1950s, by Barrett Electronics of Northbrook, Illinois, and at the time it was simply a tow truck that followed a wire in the floor instead of a rail. Out of this technology came a new type of AGV, which follows invisible UV markers on the floor instead of being towed by a chain. The first such system was deployed at the Willis Tower (formerly Sears Tower) in Chicago, Illinois to deliver mail throughout its offices.

Automated Guided Vehicles can be used in a wide range of applications and they excel in applications with the following features:

- repetitive movement of materials over a distance

- balanced processes
- regular delivery of stable loads
- processes where tracking material is important

3.1.1. Navigation

The simplest AGVs are based on **wired navigation**, where a slot is cut along the path the AGV has to follow. This wire is used to transmit a radio signal. This solution is quite invasive and not flexible. Better solutions are nowadays available on the market:

- **Guide tape:** the tapes can be one of two styles: magnetic or colored. The AGV is fitted with the appropriate guide sensor to follow the path of the tape. One major advantage of tape over wired guidance is that it can be easily removed and relocated if the course needs to change. Colored tape is initially less expensive, but lacks the advantage of being embedded in high traffic areas where the tape may become damaged or dirty.

- **Laser target navigation:** the navigation is done by mounting reflective tape on walls, poles or fixed machines. The AGV carries a laser transmitter and receiver on a rotating turret. The laser is transmitted and received by the same sensor. The angle and (sometimes) distance to any reflectors that in line of sight and in range are automatically calculated. This information is compared to the map of the reflector layout stored in the AGV's memory. This allows the navigation system to triangulate the current position of the AGV. The current position is compared to the path programmed in to the reflector layout map. The steering is adjusted accordingly to keep the AGV on track. It can then navigate to a desired target using the constantly updating position.

- **Inertial navigation:** another form of an AGV guidance is inertial navigation system (INS). An inertial navigation system is a navigation aid that uses a computer, motion sensors (accelerometers), rotation sensors (gyroscopes), and occasionally magnetic sensors (magnetometers) to continuously calculate by dead reckoning the position, the orientation, and the velocity (direction and speed of movement) of a moving object without the need for external references. It is used on vehicles such as ships, aircraft, submarines, guided missiles, and spacecraft. Other terms used to refer to inertial navigation systems or closely related devices include inertial guidance system, inertial instrument, inertial measurement unit (IMU) and many other variations.

- **Natural feature navigation**: navigation without retrofitting of the workspace is called Natural Features Navigation. One method uses one or more range-finding sensors, such as a laser range-finder, as well as gyroscopes or inertial measurement units with Monte-Carlo/Markov localization techniques to understand where it is as it dynamically plans the shortest permitted path to its goal. The advantage of such systems is that they are highly flexible for on-demand delivery to any location. They can handle failure without bringing down the entire manufacturing operation, since AGVs can plan paths around the failed device. They also are quick to install, with less down-time for the factory.

- **Vision guidance:** Vision-Guided AGVs can be installed with no modifications to the environment or infrastructure. They operate by using cameras to record features along the route, allowing the AGV to replay the route by using the recorded features to navigate. Vision-Guided AGVs use Evidence Grid technology, an application of probabilistic volumetric sensing, and was invented and initially developed by Dr. Hans Moravec at Carnegie Mellon University. The Evidence Grid technology uses probabilities of occupancy for each point in space to compensate for the uncertainty in the performance of sensors and in the environment. The primary navigation sensors are specially designed stereo cameras. The vision-guided AGV uses 360-degree images and build a 3D map, which allows to follow a trained route without human assistance or the addition of special features, landmarks or positioning systems.

- **Geoguidance:** A geoguided AGV recognizes its environment to establish its location. Without any infrastructure, the forklift equipped with geoguidance technology detects and identifies columns, racks and walls within the warehouse. Using these fixed references, it can position itself, in real time and determine its route. There are no limitations on distances to cover number of pick-up or drop-off locations. Routes are infinitely modifiable.

3.1.2 Path decision

Considering wireless AGVs, paths are pre-programmed. The AGV uses the measurements taken from the sensors and compares them to values given to them by programmers. When it approaches a decision point, it only has to decide whether to follow path 1, 2, 3, etc. This decision is rather simple since it already knows its path from its programming. This method can increase the cost of an AGV because it is required to have a team of programmers to program the AGV with the correct paths and change the paths when necessary.

3.1.3 Traffic control

Flexible manufacturing systems containing more than one AGV may require to have traffic control so the AGV's will not run into one another. Traffic control can be carried out locally or by software running on a fixed computer elsewhere in the facility. Local methods include zone control, forward sensing control, and combination control. Each method has its advantages and disadvantages.

3.1.4 Benefits & Limitations

Cost saving: cost saving comes from implementing automated guided vehicles is a reduction in labor costs because you are either replacing an existing employee or foregoing a new hire.

Safety: they can perform tasks that are dangerous to human workers, such as handling hazardous substances, working in extreme temperatures, and moving heavy materials.

Productivity: AGVs are able to operate 24/7 and in conditions that humans cannot effectively work, which ultimately increases productivity and the bottom line.

Track Inventory: when linked with a warehouse control system or a warehouse management system like Kanban, AGVs easily and automatically track inventory. It means that you know exactly how much material you have, allowing you to order materials when you need them.

Limitations: higher initial investment is a factor to be considered. Moreover, any change takes time and special skills for programming

3.2 Collaborative Robots (Cobots)

A cobot is a robot intended to physically interact with humans in a shared workspace. This is in contrast with other robots, designed to operate autonomously or with limited guidance, which is what most industrial robots were up until the decade of the 2010s.

Cobots are built with safety features such as integrated sensors, passive compliance, or overcurrent detection. The integrated sensors will feel external forces and, if this force is too high, lead the robot to stop its movement. Passive compliance is produced by mechanical components. If an external force acts on a joint, this joint will submit itself to this force. So, in the case of a collision, the joint will move in the opposite direction or stop completely to avoid causing injury.

The term "collaborative robot" is often a misnomer. In fact, although a collaborative robot is designed to work alongside humans, the device itself is not necessarily force limited. This means that the robotic cell is monitored, is safe for human co-workers, and relies on at least one of the 4 collaborative modes.

3.2.1 Collaborative modes

Safety monitored stop: if the human enters the restricted area in the pre-determined safety zone, the robot will stop all movement altogether. Notice that the robot is not shut down, but the brakes are on. An example is provided by FANUC's Dual Check Safety (DCS) control architecture. Dual Check Safety software functions (Position Check, Safe Zones, Safety Speed Check, and Cartesian Position Check) provide safety rated tools for the operator to create safety boundaries to ensure the robot doesn't move outside of restricted spaces.

Hand guiding: this type of collaboration uses regular industrial robots, but with an additional device that 'feels' the forces that the worker is applying on the robot tool. Hand Guidance allows the operator to safely control and guide the robot arm and tooling. In the past, robot movement was usually directed using the teach pendant.

Speed and separation monitoring: the environment of the robot is monitored by lasers or a vision system that tracks the position of the workers. The systems track the position of workers and adapts its speed accordingly.

Power and force limiting: this is the type of robot that everybody calls a collaborative robot. So yes, this is probably the most worker friendly robot since it can work alongside humans without any additional safety devices. They have built-in force torque sensors that detect impact and abnormal forces. The sensors stop the robot when overloaded. This means that if the robot's arm hits something (...like a worker), it automatically stops to protect its human colleagues. These features aren't present on industrial robots, and they're the reason why force limited robots can work alongside humans without any fencing. Regular industrial robots must be isolated because they neither feel nor monitor their environment.

1.2.2 Main Cobots on the market[18]

ABB IRB 1400 Yumi: this robot is specially designed to assemble small electronic devices, so it has the best repeatability out of all the collaborative robots. But there's a potential tradeoff here: with a small payload of just 0.5 kg per arm, electronic boards are basically the only thing it can handle.

---

[18] https://blog.robotiq.com/collaborative-robot-ebook

Features	Values
Degrees of freedom	7 per arm
Payload	0.5 kg per arm
Weight	38 kg
Repeatability	+/- 0.02 mm
Reach	500 mm
Safety	PL b Vat B
Price	40000 USD
Ease of programming	8/10

Table 2: ABB IRB 1400 Yumi Data Sheet

Comau Aura: at a 110 kg payload, the AURA is the cobot with the biggest payload out there. Not only does the robot have a safety skin; it has proximity and tactile sensors embedded in its skin so it can prevent impact and retract depending on the intensity of the impact.

Features	Values
Degrees of freedom	6 per arm
Payload	110 Kg
Weight	685 kg
Repeatability	+/- 0.07 mm
Reach	2210 mm
Safety	proximity and tactile sensors
Price	80000 USD
Ease of programming	6/10

Table 3: Comau Aura Data Sheet

Fanuc CR 35iA: The CR 35iA is one of the biggest collaborative robots on the market and it has a 35 kg payload. It's built over a traditional industrial robot, but its safety features make it safer than any other big robot out there.

Features	Values
Degrees of freedom	6 per arm
Payload	35 Kg
Weight	990 kg
Repeatability	+/- 0.08 mm
Reach	1813 mm
Safety	soft external skin, force torque sensor at the base of the robot: PL d Cat 3
Price	87000 USD
Ease of programming	8/10

Table 4: Fanuc CE 35iA Data Sheet

Kuka LBR IIWA 14 R820: With an excellent power to weight ratio, the LBR IIWA are equipped with highly sensitive force torque sensors at each joint. As opposed to other force limited robots that read the current in their motor, the LBR has sensors that detect micro impacts.

Features	Values
Degrees of freedom	7
Payload	14 Kg
Weight	30 kg
Repeatability	+/- 0.15 mm
Reach	820 mm
Safety	Uses Safe Operation software, Complying to ISO 10218; ISO 12100; ISO 13849
Price	70000 USD
Ease of programming	9/10

Table 5: Kuka LBR IIWA 14 R820 Data Sheet

Universal Robots UR3/U5: The UR3 is designed for polish, glue and screw applications and can be mounted on to a table station for pick-and-place or assembly in optimized production flows. The UR5 stands a bit taller, with a reach radius of 33.5 in (850 mm) and can carry payloads of up to 11 lbs (5 kg). The UR5 is best suited for pick-and-place and testing. The UR10, the largest robot of the trio, is twice as strong as the UR5, with a payload capacity of 22 lbs (10 kg).

Features	UR 3	UR 5
Degrees of freedom	6	6
Payload	3 Kg	5 Kg
Weight	11 kg	18.4 kg
Repeatability	+/- 0.1 mm	+/- 0.1 mm
Reach	500 mm	850 mm
Safety	TUV approved	TUV approved
Price	28000 USD	35000 USD
Ease of programming	8/10	8/10

Table 6: Universal Robots UR3 / U5 Data Sheets

1.2.3 Final considerations on cobots

As already mentioned, the most commonly known "cobots" apply Power and Force Limiting collaborative modes.

However, they have some limitations if compared to standard robots:

- they can be significantly more expensive
- they have limited payload capacity

On the other hand, other collaborative modes are probably more recommended for the following reasons:

- standard and less expensive robots are used
- existing robots can be transformed in cobots just by adding additional features like safety stop, limiting and separating speed monitoring functions and hand guiding hardware.

3.3 Drones

Figure 9: A drone[19]

The last category of autonomous robots worth mentioning is commonly known as drone. The term "drone" refers to an unmanned aerial vehicle (UAV), i.e. an aircraft without a human pilot on board. The term unmanned aircraft system (UAS) was adopted by the United States Department of Defense (DoD) and the United States Federal Aviation

---

[19] https://commons.wikimedia.org/wiki/File:Drone_First_Test_Flight.jpg

Administration in 2005 according to their Unmanned Aircraft System Roadmap 2005–2030[20]. In this section, we will use "drone" as main terminology.

Although drones have been developed for military scenarios, in recent years we assisted to an explosion of drones for the civilian market, which is dominated by Chinese companies. Chinese drone manufacturer DJI alone had 74% of civilian-market share in 2018, with no other company accounting for more than 5%, and with $11 billion forecast global sales in 2020[21].

Drones can have different designs, however the most common one is the quadcopter, i.e., a helicopter with four rotors. Quadcopters generally have two rotors spinning clockwise (CW) and two counterclockwise (CCW). Flight control is provided by independent variation of the speed and hence lift and torque of each rotor. Pitch and roll are controlled by varying the net centre of thrust, with yaw controlled by varying the net torque. The four-rotor design

---

[20] "Unmanned Aircraft Systems Roadmap" (PDF). Archived from the original (PDF) on 2 October 2008.

[21] Bateman, Joshua (1 September 2017). "China drone maker DJI: Alone atop the unmanned skies". News Ledge.

allows quadcopters to be relatively simple in design yet highly reliable and maneuverable. Research is continuing to increase the abilities of quadcopters by making advances in multi-craft communication, environment exploration, and maneuverability. If these developing qualities can be combined, quadcopters would be capable of advanced autonomous missions that are currently not possible with other vehicles. For small drones, quadcopters are cheaper and more durable than conventional helicopters due to their mechanical simplicity.[22]

3.3.1 Drones and Industry 4.0

In the context of Industry 4.0, drones have very similar functions as AGVs described in section 3.1, with the main obvious difference that drones fly instead of following paths on the ground. Like AGVs, drones have the capability to move components (lighter) from point A to point B, autonomously or not. Which means, a drone

- can be remotely controlled

---

[22] https://en.wikipedia.org/wiki/Quadcopter

- can be programmed to follow a specific path, or
- can fly with a high level of autonomy

Some applications of drones, especially in an industrial context, can be summarized as follows:

- **remote sensing**: drones can carry sensing equipment to assist with any number of functions. Geological surveying, agriculture, archeology, and several other industries can benefit greatly from the myriad of sensors that can be packed into a drone. For example, drones can be used in factories to monitor the health of the facility, checking the status of the roof, cranes, ducts and identifying the cause of leaking without using more expensive solutions
- **security & surveillance**: drones can be used to oversee specific areas. Can be used by Police & Security agents in a company
- **disaster relief**: the milieu of sensors that can be packed into a drone can be used to help locate and save life in the midst of natural disasters. Drones can be used to gather and deliver medical samples,

supplies, and medicine to remote or otherwise unreachable areas in a disaster zone

- **locating system**: Audi is using a specially developed drone system to locate vehicles that are ready for dispatch at the Neckarsulm site. The flying device flies over the vehicle dispatch area at the Audi site in Neckarsulm autonomously. The drone uses GPS and RFID technology to identify and save the exact position of all vehicles it flew over, thereby helping Audi employees to plan the necessary steps from completion of the vehicles to dispatch to the customers[23]
- **dispatching parts**: in a video showing the smart factory concept by Audi again, you can see how drones can be used to dispatch car components (e.g. a steering wheel) inside the assembly line[24]

---

[23]https://www.audi-mediacenter.com/en/press-releases/audi-uses-drones-to-locate-vehicles-at-neckarsulm-site-12999#:~:text=The%20specially%20developed%20hexacopter%2C%20a,of%20the%20cars%20parked%20there.

[24] https://www.youtube.com/watch?app=desktop&v=otE6CnFUXDA

Chapter 4: Big Data Analytics

4.1 Brief History 85

4.2 What Big Data are 86

4.3 Types of big data and main sources 87

4.4 Analytics of Big Data 92

4.5 Main Benefits of Big Data Analytics 99

4.6 Big Data Analytics requirements and challenges 101

Chapter Summary

"Information is the fuel of the XXI Century and Data Analytics the combustion engine"

Peter Sondergaard, Former EVP, Research & Advisory at Gartner

In this chapter, we will describe what Big Data is, what the main sources of Big Data are, how to take advantage if it with

Big Data Analytics and what the main techniques are. Finally, we will present some key requirements and challenges in exploiting Big Data Analytics properly.

Keywords: Big Data, Big Data Analytics, Data Mining, Machine Learning, Artificial Intelligence

4.1 Brief History

- **2005:** Roger Mougalas, director of market research at O'Reilly Media, coined the term Big Data. In the same year (2005), Yahoo created the now open-source Hadoop with the intention of indexing the entire World Wide Web. Today, Hadoop is used by millions of businesses to go through colossal amounts of data.

- **2009:** In the largest biometric database ever created, the Indian government stored fingerprint and iris scans of all of its citizens.

- **2010:** During a speech at the Techonomy, Eric Schmidt, Executive Chairman of Google from 2001 to 2015, presented that there were 5 exabytes of data stored since the beginning of time up to the year 2003.

- **2016:** 5 exabytes are produced every 2 days.

4.2 What Big Data are

Big Data can be described by the following characteristics:

1. **Volume:** the quantity of generated data. The size of the data determines whether it can be considered big data or not.
2. **Variety:** the type and nature of the data. This helps people who analyze it to effectively use the resulting insight. Big data draws from text, images, audio, video; plus, it completes missing pieces through data fusion.
3. **Velocity:** in this context, the speed at which the data is generated and processed to meet the demands and challenges that lie in the path of growth and development. Big data is often available in real-time. Two kinds of velocity related to Big Data are the frequency of generation and the frequency of handling, recording, and publishing.
4. **Veracity:** it refers to the data quality and the data value. The data quality of captured data can vary greatly, affecting the analysis accuracy.

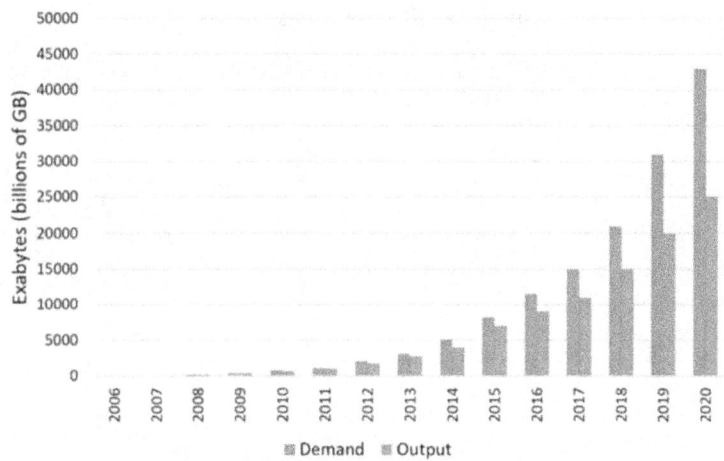

Figure 10: Storage supply and demand from 2006 to 2020[25]

4.3 Types of big data and main sources

Big Data may be classified as either structured or unstructured:

1. **Structured data:** structured data are highly organized and made up mostly of tables with rows and columns

---

[25] https://www.eetimes.com/author.asp?section_id=36&doc_id=1330462

that define their meaning. Examples are Excel spreadsheets and relational databases
2. **Unstructured data:** unstructured data is basically everything else

According to a statistic provided by EETimes, the growth of structured versus unstructured data over the past decade shows that unstructured data accounts for more than 90% of all data:

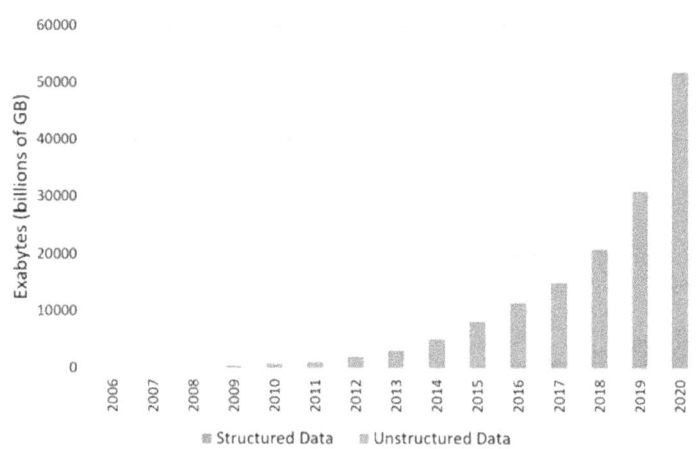

Figure 11: Explosion of digital data from 2006 to 2020[26]

Source: Patrick Cheesman

---

[26] https://www.eetimes.com/author.asp?section_id=36&doc_id=1330462

The discrepancy that the observant reader will have noted regarding the value of Exabytes per year in the previous two charts is due to the different sources.

4.3.1 Big Data main source

Big Data comes from:

- **Archives**: scanned documents, medical record, correspondence, forms, insurance etc.
- **Docs**: XLS, PDF, CSV, DOC, HTML, XML email etc.
- **Media**: Images, Video, audio, flash, podcasts, live streams etc.
- **Data Storage**: SQL, NoSQL, Cloud etc.
- **Business App**: Project Management, Productivity, ERP, MES, PLM, etc.
- **Public Web**: Government, traffic, finance, car insurance
- **Social Media**: Twitter, LinkedIn, Facebook, Google+, Instagram etc.

- **Machine Log Data**: event logs, server data, audit logs, mobile location, mobile app etc.
- **Sensor Data**: Medical devices, smart devices, car sensors, road cameras, satellite traffic recording devices, machinery, video games, assembly lines etc.

Focusing purely on the operational industrial context, Big Data comes mainly from:

1. **Cyber-Physical Systems (CPS):** Cyber-Physical Systems have been described in the IoT chapter. CPS are the essence of IoT and IIoT. There are several ways to connect physical assets, for example through RFIDs, or Ethernet, WLan and so on. Almost all solution implies to exploit digitalization. In this sense, physical systems embed cyber systems, becoming cyber-physical systems (CPS). CPSs transmit a huge amount of data in real-time, such as:
   - Logistics & asset tracking data
   - Kinematics data from robots and CNC machines
   - Temperature control data
   - Warehouse management data
   - Key Process Variable (KPV) data

2. **Point Cloud Generation Systems:** a point cloud is a set of data points in space generally produced by 3D scanners, which measure a large number of points on the external surfaces of objects around them. Point clouds are used for many purposes, including to create 3D CAD models for manufactured parts, for metrology and quality inspection, and for a multitude of visualization, animation, rendering and mass customization applications. Optical systems are nowadays quite common in industry and their application is fastly growing.

3. **Computer-Aided Engineering (CAE) Systems:** CAE is a category of computer software to aid in engineering analysis tasks. It includes finite element analysis (FEA), computational fluid dynamics (CFD), multibody dynamics (MBD), durability and optimization. CAE areas covered include:

    - Stress analysis on components and assemblies using Finite Element Analysis (FEA);
    - Thermal and fluid flow analysis Computational fluid dynamics (CFD);
    - Multibody dynamics (MBD) and Kinematics;

- Analysis tools for process simulation for operations such as casting, molding, and die press forming.
- Optimization of the product or process.

4.4 Analytics of Big Data

Big Data Analytics can be grouped into 4 main families:

- **Data Science:** data science is an umbrella of several techniques that are used for cleansing, preparation and final analysis of data. It includes data analytics, software engineering, data engineering, machine learning, predictive analytics, business analytics, and more. Unlike data mining and data machine learning, it is responsible for assessing the impact of data in a specific product or organization[27].

[27,58]https://www.datasciencecentral.com/profiles/blogs/difference-of-data-science-machine-learning-and-data-mining

- **Data Analytics:** data analytics implies the use of different technique such as descriptive statistics, data visualization data communication for conclusions. Data analysts must have a basic understanding of statistics, a very good sense of databases, the ability to create new views, and the perception to visualize the data. Data analytics can be referred to as the basic level of data science: a data scientist creates questions while a data analyst finds answers to the existing set of questions[28].

- **Data Mining:** Data mining primary goal is to extract information from various sets of data in an attempt to transform it in proper and understandable structures for eventual use. Data mining can be also seen as a confluence of various other fields like artificial intelligence, pattern recognition, visualization of data, machine learning, statistical studies and so on. While data science focuses on the science of data, data mining is concerned with the process of discovering newer patterns in big data sets.

---

[28]https://www.simplilearn.com/data-science-vs-data-analytics-vs-machine-learning-article

Unlike machine learning, algorithms are only a part of data mining.[29]

So, what is the difference between Data Analytics and Data Mining? Actually, the border is a bit blurry: we can say that Data Analytics has its roots from business analytics or business intelligence models while data mining uses more of scientific and mathematical techniques to come up with patterns and trends, although the goal is basically the same, which is to extract information from Big Data.

In Industry, statistical tools are commonly used to extract data to understand processes and improve them.

---

[29]https://www.datasciencecentral.com/profiles/blogs/difference-of-data-science-machine-learning-and-data-mining

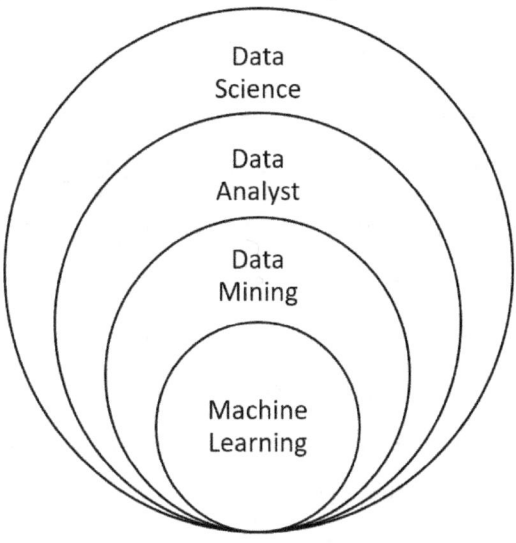

Figure 12: Levels of Data Science

Six Sigma (6σ) is a set of techniques and tools, mainly statistical, for process improvement. It was introduced in 1986 by Bill Smith, an engineer working at Motorola. In 1995 Jack Welch, CEO at General Electric, made it central to his business strategy. In a Six Sigma process, 99.99966% (+/- 6 Standard Deviation – Sigma) of parts are statistically expected to be free of defects. Six Sigma strategies seek to improve the quality of the output of a process by minimizing variability in manufacturing and business processes.

- **Machine Learning:** machine learning is kind of artificial intelligence that is responsible for providing computers the ability to learn about newer data sets without being programmed via an explicit source. In machine learning, algorithms are used for gaining knowledge from data sets. Compared to data mining, machine learning mainly focuses on algorithms. When Amazon recommends "You might also like" products, or when Netflix recommends a movie based on past behaviors, machine learning is at work.

Artificial Intelligence (AI) is intelligence demonstrated by machines, in contrast to the natural intelligence displayed by humans and other animals. AI is a branch of computer science that aims to create intelligent machines, and nowadays it has become an essential part of the technology industry. Everyone has certainly experienced AI at least once in his or her life, few examples are:

- **Siri:** Siri is the virtual assistant part of Apple Inc.'s iOS, WatchOS, macOS, HomePod, and tvOS operating systems. The assistant uses voice queries and a natural-language user interface to answer questions, make recommendations, and perform actions by delegating requests to a set of Internet services.

- **Cortana:** Cortana is a virtual assistant created by Microsoft for Windows 10, Windows 10 Mobile, Windows Phone 8.1. Invoke smart speaker, Microsoft Band, Surface Headphones, Xbox One, iOS, Android, Windows Mixed Reality, and Amazon Alexa. Cortana can set reminders, recognize natural voice without the requirement for keyboard input, and answer questions using information from the Bing search engine.
- **Computer chess:** Computer chess is a game of computer architecture encompassing hardware and software capable of playing chess autonomously without human guidance. Computer chess acts as solo entertainment (allowing players to practice and to better themselves when no sufficiently strong human opponents are available), as aids to chess analysis, for computer chess competitions, and as research to provide insights into human cognition. Current chess engines are able to defeat even the strongest human players under normal conditions. Nevertheless, it is considered unlikely that

computers could ever solve chess due to its high level of complexity.

- **Google Translate:** Google Translate uses Google Neural Machine Translation (GNMT), a neural machine translation (NMT) system developed by Google and introduced in November 2016, that uses an artificial neural network to increase fluency and accuracy in Google Translate.

Companies today use machine learning in maintenance and support services. By means of sensors, artificial intelligence helps capture the energy consumption of individual machines, analyze maintenance cycles, and then optimize them in the following stage. Operating data indicates when a part must be replaced or where there is likely to be a defect. As the amount of data increases, the system becomes better at optimizing itself and making more accurate predictions.

4.5 Main Benefits of Big Data Analytics

- **Data-driven decision making:** data mining techniques such as statistics or machine learning can extract useful information from raw data to support the decision-making process. Indeed, data-driven decision making is far more reliable and powerful than just simply opinions.

> *"It is a major mistake to theorize before one has data. One begins to alter the facts to fit them into the theories, instead of theories to fit the facts"*
> Sherlock Holmes

> *"Without Data, you are just another person with an opinion"*
> W. Edwards Deming

- **Understanding and targeting customers:** since marketing is all about reaching the right customers at the right time, Big Data Analytics can be used to predict purchases, analyze customer behaviour. The "you may also like" approach widely use now in almost all e-commerce platform is an example of consumers targeting.

- **Self-optimization:** through machine learning, systems can self-adapt, self-optimize and self-improve to achieve better performances.
- **Security:** cyber-security will have massive benefits in using data analytics techniques. IBM's fraud detection technology helped a large global money-transfer company stop more than $37 million in fraud[30]
- **Smart Factory:** Big Data Analytics is the brain of a Smart Factory: through IoT and CPS, raw data are collected (in the Cloud) and then analyzed through data mining techniques to improve performances.
- **Smart Product Development:** Smart Product Development fosters 4.0 technologies to improve all different stages in the product development process, independently from the method, approach or specific process selected. In the paper "The way from Lean Product Development (LPD) to Smart

---

[30] https://datafloq.com/read/using-big-data-to-improve-law-enforcement/3485

Product Development (SPD)"[31] the authors describe the concept of Lean Product Development as well as new requirements for an intelligent and Smart Product Development through the introduction of modern Industry 4.0 related technologies.

4.6 Big Data Analytics requirements and challenges

- **IT Infrastructure**: a high-performance IT infrastructure must be created. Managing Big Data implies transfer, store, and analyze them.
- **High-speed and reliable Internet connection**: the "always connected" state becomes absolutely a must.
- **Data Storage**: the storage capacity must be adequately provided.

---

[31] Rauch E, Dallasega P, Matt D T, The way from Lean Product Development (LPD) to Smart Product Development (SPD), 26th CIRP Design Conference, Procedia CIRP 50 (2016) 26 – 31

- **Computing Power**: the number of operations per unit of time. Big Data requires by high computing power.
- **Cybersecurity**: this is a recurring topic throughout the entire book. We met cyber-security talking about Internet of Things and Cloud Storage, in which data are collected and then stored to be further analyzed.
- **Skilled data scientist**: first of all, being a data scientist implies having a high-level education (Master or PhD) in technical fields. The most common fields of study are Mathematics and Statistics (32%), followed by Computer Science (19%) and Engineering (16%). A degree in any of these courses will give you the skills you need to process and analyze big data[32].

Another important aspect is the knowledge of some of the most powerful analytical tools:

- **Python Coding**: it is the most common coding language required in data science roles, along with Java, Perl, or C/C++. Python is a great programming language for data scientists. Because of its versatility, you can use Python for

---

[32] https://www.kdnuggets.com/2018/05/simplilearn-9-must-have-skills-data-scientist.html

almost all the steps involved in data science processes. It can take various formats of data and you can easily import SQL tables into your code. It allows you to create datasets and you can literally find any type of dataset you need on Google.

- **SQL Database Coding**: even though NoSQL and Hadoop have become a large component of data science, it is still expected that a candidate will be able to write and execute complex queries in SQL. SQL (structured query language) is a programming language that can help you to carry out operations like add, delete and extract data from a database. It can also help you to carry out analytical functions and transform database structures.

- **Apache Spark**: Apache Spark is becoming the most popular big data technology worldwide. Apache Spark is specifically designed for data science to help run its complicated algorithm faster. It helps in disseminating data processing when you are dealing with a big sea of data thereby, saving time. It also helps data scientist to handle complex unstructured data sets. You can use it on one machine or cluster of machines.

Chapter 5: The Cloud

5.1 Brief History 107

5.2 Benefits 108

5.3 Limitations 111

5.4 Service Models 113

5.5 Industrial applications of Cloud Computing 126

Chapter Summary

The Cloud (or Cloud Computing) *"is where software applications, data storage, processing power and even artificial intelligence are accessed over the Internet from any kind of computing device"*. In practical terms, the cloud is made up loads of giant data centres – also known as "server farms" – run by Google, Amazon, Microsoft, IBM, Apple and a host of other traditional and emerging computing giants.

In this chapter, we will describe what the main benefits and limitations of using cloud solutions are and main service models. Finally, industrial applications of cloud computing will be presented.

Keywords: Cloud, Cloud Computing, SaaS, PaaS, IaaS

If you are using popular applications such as Google Docs to create documents, Dropbox to share big files, iCloud to back up your iPhone or Hotmail to check your mailbox, in most cases you are not using a software installed in your PC, but probably a web browser is the only installed software you need. Moreover, data are physically stored in some server somewhere owned by a provider, but it is in most cases irrelevant. This is probably one of the main reasons why cloud computing is so scary and resisted in many corporate data centres. But on the other hand, it is also why cloud computing is so powerful for the vast majority of applications.

5.1 Brief History

- **2006:** Amazon popularized the term "cloud computing" and created subsidiary Amazon Web Services (AWS). AWS is a subsidiary of Amazon that provides on-demand cloud computing platforms to individuals, companies and governments, on a paid subscription basis. The technology allows subscribers to have at their disposal a virtual cluster of computers, available all the time, through the Internet.

- **2008:** Google released Google App Engine (GAE) in beta. GAE is a web framework and cloud computing platform for developing and hosting web applications in Google-managed data centers.
- **2010:** Microsoft released Microsoft Azure, a cloud computing service created by Microsoft for building, testing, deploying, and managing applications and services through a global network of Microsoft-managed data centers.

- **2012:** Oracle announced the Oracle Cloud. This cloud offering is poised to be the first to provide users with access to an integrated set of IT solutions, including the Applications (SaaS), Platform (PaaS), and Infrastructure (IaaS) layers.

- **2013:** Google Compute Engine was released. Google Compute Engine is the Infrastructure as a Service component of Google Cloud Platform which is built on the global infrastructure that runs Google's search engine, Gmail, YouTube and other services. Google Compute Engine enables users to launch virtual machines on demand.

5.2 Benefits

- **Security, Privacy and Reliability:** *"From the perspective of a hacker... it is infinitely easier for me to break through the meagre security on a personal computer than it is for me* *to take on a Google server"*[33]. Security issue may be accounted for contractually. For example, in October 2009 Los Angeles City Council decided to move its 30000 employees to Google's Government Cloud services and a security-breach penalty clause was added to the contract.

- **Device Independent:** data and applications are accessible from any connected computer. If you are working on a project using Google docs, for example, you don't have to worry if the computer on which you are working contains the most recent version of the file. If you lose your laptop or if your hard disk brakes, your work is still in the cloud, and you can continue to work on another device.
- **High collaborative:** co-workers are always sure they are working on the latest version. It doesn't matter if they are all based in different places, different countries or continents, as long as they are connected, they can always collaborate on the same updated file.

---

[33] Barnatt C, A Brief Guide to Cloud Computing: An essential guide to the next computing revolution. (Brief Histories) Little, Brown Book Group. Kindle Edition, 2010

- **Task centric:** the cloud is focused around what users want to achieve, rather than any particular software, hardware or network infrastructure.
- **Dynamically scalable:** any user can draw as many or as few computing resources from the cloud as they require at any particular moment.
- **Democratization:** *"Cloud computing can deliver CAE to the masses with minimum investment. Start-ups or small companies can access an incredible deal of computing power to run demanding simulations."* –Dominique Lefebvre, head of Product Management at ESI Group
- **Cost savings:** in most cases, cloud solutions are free, nobody is paying or is expecting to pay to open a mail account or to use Google Docs. The Telegraph Media Group, for example, expects to reduce its software costs by 80% over 3 years following its switch from local Microsoft software to Google Apps
- **No fixed costs:** traditionally, computing involves substantial fixed costs, which includes the cost of building, equipping and maintaining data centres, as well as software licenses and customer support. Cloud computing is dynamically scalable and task-centric, and for most

users it has no fixed costs. Rather, all costs are on a per-usage or variable basis.

- **Environmentally friendly:** the world's data centres already have about the same carbon footprint as the airline industry. Compared to in-house or desktop computing, cloud computing uses server providers which can run their infrastructure more efficiently. With about half of the energy used by a large data centre going into cooling, putting cloud server farms in very cold countries would be a wise choice, saving money and carbon footprint. In preparation for its anticipated cloud computing "cold rush", Iceland is laying high-capacity, fiber optic cables to connect the country with North America and Europe.

5.3 Limitations

- **Reliable and high-speed internet connection:** all the benefits listed in the previous section have no meaning if an internet connection is not available. Moreover, high-speed connection is in most cases required where power computing is demanding.

- **Regulations:** in some specific situations, regulations may limit the use of Cloud Computing. For instance, Export Control Regulations. Export control regulations are international laws that prohibit the unlicensed export of certain commodities or information for reasons of national security or protections of trade. Export controls usually arise for one or more of the following reasons:
  - The nature of the export has actual or potential military applications or economic protection issues.
  - Government concerns about the destination country, organization, or individual.
  - Government concerns about the declared or suspected end use or the end user of the export. Export Control plays a role as soon as sensitive documentation is exported in another country, including sending email through servers located in different countries.
- **Cybersecurity:** although it is true that the Cloud can be considered a safe environment for most of the application, we can't say the Cloud in 100% safe. There is

still a small probability that a cyberattack will be successful. Therefore, for all these situations where high-confidential information is shared, Cloud can be still risky. In Business, high-confidential information may include Intellectual Property (IP). In this case, a separate storage (such as portable hard disk or pen drive) is still probably the best option.

5.4 Service Models

There are essentially 3 ways in which a business may replace traditional in-house systems with cloud computing (see Figure 13):

- Software (and Storage) as a Service (SaaS)
- Platform as a Service (PaaS)
- Infrastructure as a Service (IaaS)

All involve a cloud vendor supplying servers on which their customers store data and run applications.

When companies choose SaaS, they can only run those applications that their supplier has to offer. If they opt for PaaS, they can create their own applications but only in a way determined by their cloud supplier. And last, when they opt for IaaS, they can run any applications they want on cloud hardware of their own choice.

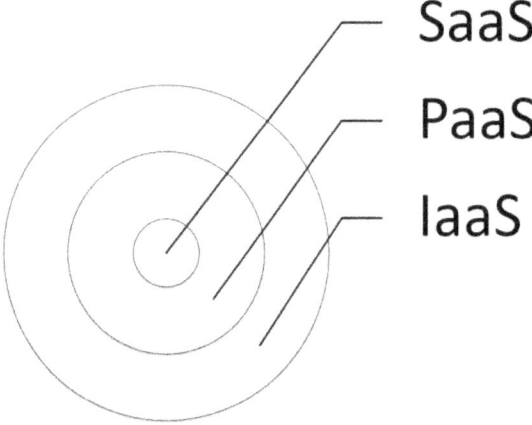

Figure 13: Service models in Cloud Computing

5.4.1 Software as a Service (SaaS)

Software as a Service (SaaS) *"is the technical term for a computer application that is accessed over the Internet instead of being installed on a computer or in a local data centre"*[34]. As already mentioned, one of the main benefits of SaaS applications is to be accessible from any device, including home PC, work PC, netbook, tablet or smartphone.

Applications include:

- **SaaS E-Mail:** most of you probably now use free on-line e-mail service, such as Gmail, Yahoo! Mail or Windows Live Hotmail. Another option is to install an e-mail application, such as MS Outlook. When people use on-line services, their messages never leave the cloud, and the e-mail software used to write and read the message is also never installed on user's PC.
- **Office SaaS:** almost everyone needs traditional office application such as a word processor or a spreadsheet, which are also available as a cloud-based office package, typically for free. All the following examples are Microsoft Office compatible:

---

[34] Barnatt C, A Brief Guide to Cloud Computing: An essential guide to the next computing revolution. (Brief Histories) Little, Brown Book Group. Kindle Edition, 2010

- **Google Docs**: the suite consists of a word processor, spreadsheet and PowerPoint presentation application, coupled with an online storage service
- **Google Suite**: it includes Gmail, Google Calendar, Google Video, the Google Sites website and intranet creation tool and loads of storage space
- **Zoho**: it contains word processing, spreadsheets, presentations, databases, note-taking, wikis, web conferencing, customer relationship management, project management, invoicing, and other applications
- **Acrobat**: Adobe created its own suite which includes pdf creator, a cloud-based file sharing and storage, a web conferencing tool called ConnectNow, a word processor (Buzzword), presentations package (Presentations) and a spreadsheet (Tables)
- **SlideRocket**: it is an online presentation package for high-professional presentations and far more sophisticated than any of its SaaS competitors

- **SaaS Desktop:** some companies now deliver cloud-based operating system and a desktop on which run them:

- **ZeroPC**: it provides a Windows-like desktop loaded with a word processor, spreadsheet, presentations package and other applications, including, for example, a pdf reader, media player and messaging application
- **EyeOS**: it is an opensource SaaS Desktop
- **IT Farm**: it provides a paid service where standard Microsoft Office and other Windows Packages are installed on their servers and delivered via a web browser. It is like running a cloud-based copy of Windows

- **SaaS Photo and Video Editing**
    - **Photoshop Express**: it is an on-line version of the Photoshop image editing package. Registration is for free;
    - **Pixlr**: in contract to Photoshop Express, Pixlr looks and behaves like the traditional full version of Photoshop running in a web browser

- **SaaS Business Applications**

Most of the SaaS applications I have listed so far cover almost all needs for home users, but to run a business something different maybe is required:

- **Zoho Creator**: this application has been previously introduced when Office SaaS was described. Zoho has also a business software application called Zoho Creator. It can be thought of as a web-based version of Microsoft Access, but from a certain point of view even more powerful because it offers the ability to create online databases that can be used simultaneously by many company's employees or customers.
- **Salesforce**: Salesforce CRM Sales is a popular sales management application
- **Employease**: it provides human resource information systems that allow companies to run their payroll, benefits administration and other personnel-related IT in the cloud
- **Clarizen**: it describes itself as providing "online work management software for companies". It offers project-management application that can be

collaboratively used to manage resources, timesheets, budgets or expenses

- **Netsuite**: it claims to be "the first and only online business application to support an entire company". It offers tools for Customer Relationship Management (CRM), Enterprise Resource Planning (ERP), e-commerce and website management.
- **WebEx WebOffice**: it is a suite of online collaboration tools. The applications allow people to hold virtual meetings by using phone conferencing, share documents and share desktops

5.4.2 Storage as a Service (SaaS)

SaaS has the same acronym as Software as a Service. It provides the facility to store, share and back up files on the Cloud. Popular services include:

- https://www.box.com
- https://www2.livedrive.com/
- https://www.dropbox.com/

- https://www.icloud.com/
- https://onedrive.live.com
- https://www.google.com/drive/

5.4.3 Platform as a Service (PaaS)

Platform as a Service enables to access online hardware in a software environment in which companies can develop and run their own SaaS applications. PaaS vendors provide everything necessary to create, test and deliver new online applications. Possibilities ranges from the development of new business systems for a particular organization, through the development of online customer interfaces using PaaS tools to help bring new SaaS applications to market. For example, Google offers a PaaS service called App Engine. App Engine allows anybody to write new cloud applications and to deliver them over the web using Google's infrastructure.

One of the major advantages of many PaaS offerings is that new applications do not have to be migrated between systems. It is therefore unlikely that a new application that has been properly tested will not work perfectly the moment it goes live on the web. However, PaaS is not a perfect solution. One of the biggest potential drawbacks of developing new applications using PaaS is vendor lock-in. There are currently a relatively small number of PaaS vendors available, and all of them have their own standards and programming tools. Therefore, users inevitably become reliant on their chosen vendor, making the choice of vendor critical, and is likely to drive most users to the doors of large organizations like Google and Microsoft.

Other PaaS providers are:

- **Force**: this is the PaaS offer from Saleforce.com. It allows anybody to build and run applications on the same infrastructure used for Salesforce's off-the-shelf SaaS applications
- **Microsoft Windows Azure**: Azure is Microsoft's platform for running Windows applications and storing data in the cloud.

Technically speaking, also online website builders fall into the PaaS category:

- Google Sites

- Moonfruit
- Wix
- Webs
- WordPress

5.4.4 Infrastructure as a Service (IaaS)

"IaaS is where a vendor offers computer hardware in the cloud on which their customers can store data and develop and run whatever applications they please. IaaS therefore allows companies to move their existing programs and data into the cloud and to close down their own local servers and data centres"[35]

The fundamental building block of computing infrastructure is the server, which can be defined as a piece of hardware that offers remote processing power and/or storage capacity. Servers that can be accessed over the Internet are therefore what IaaS vendors supply to their customers.

---

[35] Barnatt C, A Brief Guide to Cloud Computing: An essential guide to the next computing revolution. (Brief Histories) Little, Brown Book Group. Kindle Edition, 2010

Servers are expensive hardware for the following reasons:

- constantly need power supply;
- need great deal of cooling;
- need a secure environment to protect them from fire or other natural disaster;
- must be backed up regularly;
- need 24/7 IT support to ensure they work properly and to effect repairs if necessary.

For these reasons, IaaS is a convenient solution for SME, startups and any business that can't afford a data centre. On the other hand, cloud computing servers must be connected with robust, high-speed connection. Although few years ago servers could be associated with a physically discrete hardware box, today it is not true anymore. Indeed, most servers consists server blades built in racks. A rack is an equipment stand, usually around 50cm wide, on which a number of server computers can be stored. Today the racks in most large data centres contain blade servers. Basically, a blade server is a computer circuit board with a processor, memory and a hard drive or two attached. The benefit of blade servers is that they save a great deal of space and use less energy because each server blade does not need its own, individual power supply. Nowadays modern racks can accommodate up to 128 server blades.

IaaS vendors may rent virtual or real servers, which means that while cloud data centres contain rack upon rack of server blades, these physical servers are subdivided by a software process called "virtualization".

Figure 14: Server Racks[36]

---

[36] https://commons.wikimedia.org/wiki/File:Half_filled_server_racks.jpg

There are currently 4 categories of IaaS services:

- **Private Cloud**: this is potentially the most secure form of IaaS. Physical servers are all located in the same part of a data centre, which means that their cloud hardware is as separated as possible from that of other users
- **Dedicated Hosting**: a customer rent a number of dedicated physical servers within a cloud data centre. Like private cloud, the customer doesn't share their hardware with anybody else, however the customer doesn't have control of the location of its physical server. A benefit of this approach is that it can be dynamically scaled: when the customer needs to increase or decrease the number of servers they are using, they can do that easily even on hourly basis
- **Hybrid hosting**: a customer rents dedicated physical servers, but with virtual servers' instances added into the mix to increase flexibility at minimum cost
- **Cloud Hosting**: a customer can buy as many virtual server instances s required on demand, but there is no control where data are stored and their applications run. They simply share server blades with other customers.

Although for some companies this is too risky, cloud hosting is without doubt the most technically and environmentally efficient form of cloud computing. Several companies are now operating in the IaaS marketplace:

- **Amazon Web Services (AWS)**: Amazon's IaaS is fully virtualized, with the company selling virtual server instances
- **GoGrid**: it offers cloud hosting, hybrid hosting and dedicated hosting solutions
- **Rackspace**: it is a well-known provider of traditional hosting services

5.5 Industrial applications of Cloud Computing

In the factory of the future, physical assets like machines, robots, fixture and so on communicate each other and share information at all company levels by means of the cloud. We can imagine a virtual invisible cloud full of data to be shared and analyzed through AI to improve flexibility, quality, efficiency and effectiveness of the production system. Following, some examples of industrial applications are be presented:

- **Work instructions:** manual operations are typically supported by work instructions. "Paperless" supports are becoming more popular for several reasons:
  - paper can be lost, while digital are always stored safely
  - digital work instructions are always up to date
  - they are environmentally friendly, as the use of paper is eliminated
  - videos, links and data input can be used
  - they can be automatically translated in different languages

Moreover, the use of work instructions in the Cloud adds even more benefits:

- they can be shared throughout different plants, located in different locations
- the revision alignment is guaranteed

- **Digital platforms:** in December 2016, in the Business section of The Economist the article "Siemens and General Electric gear up for the internet of things" was published. The article deals with two the two similar approaches that these two big companies are using to face up to the Digital Transformation. The answers are called Predix for GE and Mindsphere for Siemens: they are open cloud platforms or "IoT operating systems" for applications in the context of the Internet of Things. Operational data are stored and made accessible through digital applications to allow industrial customers to make decisions based on valuable factual information. As cloud-based PaaS (platform as a service), they collect and analyze all kinds of sensor data in real time. This information can be used to optimize products, production assets and manufacturing processes along the entire value chain.

- **Virtual teams:** virtual teams are made by people located in different buildings, cities, countries or continents. They may be part of the same organization or they may come from different companies. Virtual teams communicate through digital technologies, using the cloud.

Chapter 6: Cyber-security

6.1 Brief History 131

6.2 Top 5 most notorious cyber-attacks 132

6.3 Basic concepts 136

6.4 Defense methods 142

Chapter Summary

"If you are not concerned about cybersecurity, you don't know enough about it."[37]

Cyber-crime is now the fastest-growing industry on the planet, with estimated revenues of $445 billion in 2016, according the World Economic Forum.

---

[37] Meeuwisse R, "Cybersecurity for Beginners", Cyber Simplicity Ltd; 2nd edition (March 14, 2017)

In this chapter, we will present some notorious cyber-attacks in history and we will introduce you to basic principles of cybersecurity. In the last section, we will provide you some practical defense methods against cyber-attacks.

Keywords: cyber-security, cyber-attacks, malwares

6.1 Brief History

- **1983:** Cyber-security begins together with Internet: as soon as PC started to be connected each other, hacking was possible. However, at the dawn on internet connection, the speed was much slower than it is today and it costed a fortune. Moreover, the probability to be infected by viruses was fairly small, and the maximum consequence was just to re-install files. The key factor that gave rise to cybersecurity threats was that Internet connection speeds became faster, cheaper and more widely adopted. This change, together with faster computer processing speeds and better web application programming, gradually made it easier, more effective and cheaper to provide mainstream services through the Internet, rather than using traditional offline routes.

- **2005:** Until this date, the IT hardware and software were controlled by the company IT department and technologies operated almost exclusively within the internal company network. Then, the Cloud arrived: indeed, in chapter 9 we learnt that Amazon launched its Cloud service (AWS) in 2006. Apple with their iPhone and App Store contributed with the change in thinking. It was clear for companies that, applying a similar philosophy, they would have more choices, greater flexibility and lower costs. This change in scenarios drastically changed the role of IT technologists: they will focus more on cybersecurity to guarantee safety and reliability of IT tools, reducing vulnerability that might be leveraged (vulnerability is "a weakness that could be compromised and result in damage or harm").

6.2 Top 5 most notorious cyber-attacks

- **WannaCry**: the WannaCry attack put ransomware, and computer malware in general, on everyone's map, even those who don't know a byte from a bite. The four-day WannaCry epidemic knocked out more than 200,000 computers in 150 countries. This included critical infrastructure. In some hospitals, WannaCry encrypted all devices, including medical equipment, and some factories were forced to stop production.

- **NotPetya / ExPetr:** the worm moved around the Web, irreversibly encrypting everything in its path. Although it was smaller in terms of total number of infected machines, the NotPetya epidemic targeted mainly businesses, partly because one of the initial propagation vectors was through the financial software MeDoc. The damage from the NotPetya cyberattack is **estimated at $10 billion**, whereas WannaCry, according to various estimates, lies in the $4–$8 billion range.
- **Stuxnet:** nothing could match Stuxnet for complexity or cunning — the worm was able to spread imperceptibly through USB flash drives, penetrating even computers that were not connected to the Internet or a local network. The worm manifested itself only on computers operated by Siemens programmable controllers and software. On landing on such a machine, it reprogrammed these controllers. Then, by setting the rotational speed of the uranium-enrichment centrifuges too high, it physically destroyed them.

- **DarkHotel:** on connecting to a hotel network, they were prompted to install a seemingly legitimate update for a popular piece of software, and immediately their devices were infected with the DarkHotel spyware, which the attackers specifically introduced into the network a few days before their arrival and removed a few days after. The stealthy spyware allowed the cybercriminals to conduct targeted phishing attacks.

- **Mirai**: devices whose security had never been considered and for which no antiviruses existed suddenly began to be infected on a massive scale. These devices then tracked down others of the same kind, and promptly passed on the contagion. This zombie armada, built on a piece of malware romantically named Mirai (translated from Japanese as "future"), grew and grew, all the while waiting for instructions. Then one day — October 21, 2016 — the owners of this giant botnet decided to test its capabilities by causing its millions of digital video recorders, routers, IP cameras, and other "smart" equipment

In 2014, Ofcom[38] reported that average UK adult spent more time per day using digital devices (8h and 41 min) than sleeping (8h and 21 min). The main reason is that smart devices make people more powerful: nowadays it is almost impossible to get lost by using google maps, it is simple to check and book restaurants nearby, check and make transactions on bank account, find information easily googling everything and so on. Now a personal question: are you aware of the type and amount of information that your web accounts save? I recently discover it by myself setting my Google account and I was astonished. So, if you are curious, please have a look. The question is: are these technologies 100% secure?

---

[38]https://www.ofcom.org.uk/__data/assets/pdf_file/0031/19498/2014_uk_cmr.pdf

6.3 Basic concepts

In his book "Cybersecurity for Beginners" [39], Meeuwisse describes the steps to build an effective cybersecurity system:

- **Identify your valuable asset:** to apply the proper level of security, it is necessary to understand that the value of different types of information determines how much protection it requires. Classifying our information lets us know what to defend, but we still need to understand where to defend it. The cyber defense points are the digital locations where we could add cybersecurity controls. Typically, 6 layers of digital defense points are identified:
    - **Data**: any digital information
    - **Devices**: computers, smartphones, tablets, USB drives are just some examples
    - **Applications**: any programs/software installed in a device

---

[39] Meeuwisse R, "Cybersecurity for Beginners", Cyber Simplicity Ltd; 2nd edition (March 14, 2017)

- **Systems**: group of applications that work together to serve a more complex purpose
- **Networks**: a group of devices that are connected together physically (by wires) or virtually (using applications)
- **Other communication channels:** CD-ROM drives, USB ports, wi-fi, Bluetooth etc.
- **Protect with appropriate security**: 4 major categories of security controls can be identified. The following list ranks them form the safer to the less effective:
  - **Physical**: put information to restricted equipment, physically separated. Examples are digital memory card, a USB drive and so on
  - **Technical**: the use of a digital method to command how something can or cannot be used. Removing the ability to cut or paste information on a smartphone is an example of a technical control that can be used to minimize security risks
  - **Procedural**: instructions during a sequence of required steps to limit how something is or is not permitted to be used. An example is to require a

minimum of 2 authorized people to approve any access
- **Legal**: the use of legislation to help promote and invest in positive security methods and also to deter, punish and correct infringements.

- **Detect, respond, recover:** it implies the detection of any compromised account or device, quarantine the problem and identify countermeasures and recover by replace, restore or fix compromised assets. There are 3 main control modes to protect a digital device:
  - **Preventive controls**: it protects the device before an event happens
  - **Detective controls**: it monitors and alerts me if something happens
  - **Corrective controls**: it rectifies any gaps after the problem has been identified

6.3.1 Human factors

"People are regarded as the weakest link in cybersecurity"[40]

The most significant human factors are:

- **Inadequate cybersecurity knowledge**: cybersecurity is not a static discipline and an ongoing and substantial personal investment is required to stay up to date
- **Poor capture and communication of risks**: people tend to notice but not to report risks (culture and relationship issues)
- **Under-investment in security training**: it results in a low level of awareness and management
- **Using trust instead of procedures**
- **Absence of a single point of accountability**, the principle that all critical assets, processes and actions must have clear ownership and traceability to a single person

---

[40] Meeuwisse R, "Cybersecurity for Beginners", Cyber Simplicity Ltd; 2nd edition (March 14, 2017)

- **Social Engineering** is the art of manipulating people through personal interaction to gain unauthorized access to something

6.3.2 Threats

What are main threats and how they can attack us? In the cyber-security context, main threats are malwares and they can enter in our systems through the attack surface. The attack surface is the sum of the different points where an unauthorized user ("attacker") can try to enter data into or extract data from an environment. The six layers of digital defense points are the attack surface.

Malwares: a malware is a malicious software, such as a virus, which is specially designed to disrupt or damage a computer system. A virus is a form of malware that spreads by infecting to other files and usually seeks opportunities to continue that pattern. Malwares are typically used to:

- infect, which means to create damage or disruption
- steal information

- take instruction from the attacker
- steal confidential information
- steal money from a bank account
- block the user's access until a ransom is paid (ransomware)

6.3.3 Types of malwares

- Phishing is using an electronic communication (e.g. email) that pretends to come from a legitimate source, in an attempt to get sensitive information (e.g. a password or a credit card number) from the recipient or to install a malware on the recipient's device
- Spear phishing is a more evolute form of phishing. It describes the use of an electronic communication that targets a particular person or group of people and pretends to come from a legitimate source
- Polymorphic malwares are malicious software that can change its attributes to help avoid detection by anti-malware

6.4 Defense methods[41]

6.4.1 Basic

- **Install an effective anti-malware:** a computer program designed to look for specific files and behaviors that indicate the presence or the attempted installation of malicious software
- **Install a Firewall**: hardware or software used to monitor and protect inbound and outbound data by applying a set of rules (the firewall policy)
- **User Access Control**: rules and techniques used to manage and restrict entry to or exit from a physical, virtual or digital area through the use of permissions
- **Data retention and destruction**: a non-technical strategy which means to destroy data following certain criteria. An easy example of this principle is email: after a certain period of time, emails will be deleted
- **Password Management**

---

[41] Meeuwisse R, "Cybersecurity for Beginners", Cyber Simplicity Ltd; 2nd edition (March 14, 2017)

1. **Use strong passwords**: according to security experts, using a chain of 4 or 5 random words in lower case letters is thousands of times more difficult to crack than a password with 8 characters that contains combinations of numbers, special characters and upper- and lower-case letters. This because password-cracking software are educated about human password patterns: password guessing software can start by expecting that the first character is a capital letter and the last character is most likely to be a special character or set of numbers,
2. **Change Passwords regularly**
3. **Access to every single online service with a different username and password.** Indeed, almost every service requires its own username and password, and instead of using different usernames and passwords, many people follow the unsafe practice of re-using the same ones. This means that instead of cracking passwords, criminals can process long lists of historic usernames and passwords into automated software that can check if they work in thousands of different online services.

- **Create high-security zones**: create high-security zones that add substantially more protection to the most sensitive data. Sometimes closed system are used (not connected with a public network)
- **Implement segmentation**: to reduce security risk, a method is to subdivide the attack surface using network segmentation, which means splitting a collection of devices and applications that connect, carry or safeguard data into smaller sections. This allows for more discrete management of each section, allowing greater security to be applied in sections with the highest value, and also permitting smaller sections to be impacted in the event of a malware infection or other disruptive event
- **Timely patch management**: A controlled process used to deploy critical, interim updates to software on digital devices
- **Regular backup**: Regular back-up is a key process that can allow computer systems to be restored in the event of a successful attack.
- **Establish a Computer User Policy**: Educate users to a good security practice

6.4.2 Advanced

- **Multi-factor authentication:** this means using more than one form of proof to confirm the identity of a person or device attempting to request access (fingerprint, face recognition)
- **Encryption/Cryptography**: the art of encoding messages so that they cannot be read by anybody who intercepts them
- **Proxy servers**: programs used to provide intermediate services between a requested transaction and its destination. Proxy servers enhance security by hiding exact information about locations and users in a particular network. Attackers frequently use proxy servers for the same goal
- **Penetration testing**: it checks and scans on any application, system or website to identify any potential security gap (vulnerabilities) that could be exploited. The process can involve hackers that a company pays to manually try to identify security weaknesses. There are also automated tools that can perform similar assessments

- **Vulnerability assessment**: the identification of security gaps in a computer, software application, network or other section of a digital landscape

Create Honeypots: they are electronic devices or collection of data that are designed to trap would-be attackers by detecting, deflecting or otherwise counteracting their efforts. The honeypot will contain nothing of real value to the attacker, but will contain tools to identify, isolate and trace any intrusion

Chapter 7: Internet of Things (IoT)

7.1 Introduction to Internet of Things 149

7.2 Brief History 153

7.3 Industrial Internet of Things (IIoT) 154

7.4 Cyber-Physical Systems (CPS) 155

7.5 Establishing a communication 157

7.6 IoT Protocol & Standards 159

7.7 Applications of IoT 160

Chapter Summary

Internet of Things is probably the most popular concept associated with Industry 4.0. In the first chapter, we have seen that Internet of Things is the primary technology to be exploited, according to Henning Kagermann, Wolf-Dieter Lukas and Wolfgang Wahlster. In this chapter, we will describe Internet of Things in more detail, we will see how IoT enables the development of cyber-physical-systems and we will present what the main challenges are. Finally, practical applications of Industrial Internet of Things (IIoT) will be shown.

Keywords: Internet of Things, IoT, Industrial Internet of Things, IIoT, Cyber-Physical-Systems

7.1 Introduction to Internet of Things

The Internet of things (IoT) is the network of devices that contain electronics, software, actuators, and connectivity which allows these things to connect, interact and exchange data. IoT involves extending Internet connectivity beyond standard devices to any range of traditionally non-internet-enabled devices and everyday objects.

In the article published by the Economist "Siemens and General Electric gear up for the internet of things", it is pointed out that linking the physical and the digital worlds via the IoT could create up to $11 trillion in economic value annually by 2025, estimates the McKinsey Global Institute. A third of that could be in manufacturing[42].

In 2019, Engineering.com conducted research sponsored by Siemens PLM Software, regarding the importance of IoT in

---

[42] Siemens and General Electric gear up for the internet of things, https://www.economist.com/business/2016/12/03/siemens-and-general-electric-gear-up-for-the-internet-of-things , 3 Dec, 2016

Product Development[43]: 234 product development professionals have ben surveyed about their company's intentions regarding adding IoT functionality to their products. The respondents came from a wide range of industries, representing processed goods, component and part manufacturers, and finished goods OEMs. Here are a few highlights:

- IoT functionality in product development became more prevalent
- Respondents who held roles of manager and above generally considered implementing IoT enabled features to be more important than other members of a product development team
- Teams who are evaluating IoT enabled products are exploring a wide range of business models and product features

The importance of interconnected systems can be summarized in 4 concepts:

- data-driven decision making

---

[43] Research Report: The State of IoT Adoption in Product Development 2019, Engineering.com, 2019

- communication
- Just in Time
- Adaptability

Data-driven decision making: it means making decisions that are backed up by hard data rather than making decisions that are intuitive or based on observation alone. As business technology has advanced exponentially in recent years, data-driven decision making has become a much more fundamental part of all sorts of industries, including important fields like medicine, transportation and equipment manufacturing.

Communication: the importance of communication between physical assets can be described as follow:

- Data from physical assets promotes motivation by informing and clarifying the employees about the task to be done
- Communication improves the decision-making process, as stated above
- Data assist in controlling processes. Sharing information improve the level of understanding at all company's level,

both in vertical (hierarchic) and horizontal (equal), improving transparency

Just in Time: JIT is a methodology aimed primarily at reducing times within production system as well as response times from suppliers and to customers. Its origin and development were in Japan, largely in the 1960s and 1970s at Toyota. Since within IoT systems communicate and share data in real-time, there is no waste of time between the sending and the receiving of the information. Moreover, IoT implies the use of other 4.0 technologies like AI and autonomous robots so actions can be taken autonomously with almost no additional time required.

Adaptability: the downstream operations are influenced by upstream processes. Therefore, upstream changes mean effectively that downstream processes need to adapt themselves to new operating conditions. Techniques like Six Sigma, for example, aim to produce long-term defect levels below 3.4 defects per million opportunities (DPMO). However, especially in the context of high-personalization production, where products variation is increasingly higher, to achieve this target is simply not cost effective. In this sense, IoT technologies will create a new process paradigm, where flexible and adaptable process will deliver high-quality products with less human effort.

7.2 Brief History

- **1982**: the concept of a network of smart devices was discussed as early as 1982, with a modified Coke machine at Carnegie Mellon University becoming the first Internet-connected appliance able to report its inventory and whether newly loaded drinks were cold.

- **1991**: Mark Weiser's paper on ubiquitous computing, "The Computer of the 21st Century", as well as academic venues such as UbiComp and PerCom produced the contemporary vision of IoT. In 1994, Reza Raji described the concept in as "[moving] small packets of data to a large set of nodes, so as to integrate and automate everything from home appliances to entire factories".

- **1999**: the term "Internet of things" was likely coined by Kevin Ashton of Procter & Gamble, later MIT's Auto-ID Center, though he prefers the phrase "Internet for things". He viewed Radio-frequency identification (RFID) as essential to the Internet of things.

7.3 Industrial Internet of Things (IIoT)

IIoT is simply the application of IoT to industrial applications, including, but not limited to, manufacturing and energy management. Despite IoT, IIoT history begins earlier, in 1968, with the invention of the programmable logic controller (PLC) by Dick Morley, a General Motors employee. With the introduction of Ethernet in 1980, people began to explore the concept of a network of smart devices, and IoT era then started with the Coke machine, as explained before.

One of the first consequences of implementing the IIoT would be to create instant and ceaseless inventory control. Another benefit of implementing an IIoT system is the ability to create a digital twin of the system. Utilizing this digital twin allows for further optimization of the system by allowing for experimentation with new data from the cloud without having to halt production or sacrifice safety, as the new processes can be refined virtually until they are ready to be implemented. A digital twin can also serve as a training ground for new employees who won't have to worry about real impacts to the live system.

7.4 Cyber-Physical Systems (CPS)

CPS are the essence of IoT and IIoT. Essentially, CPS are physical assets that are able to communicate and share information by using "cyber" technologies.

These cyber-technologies can be grouped as follow:

- **Embedded controllers:** they consist of small computers that allow engineers to add "intelligence" to products at a relatively low cost.
- **Sensors:** a wide range of sensors can be used to monitor systems like accelerometers, gyroscopes,

temperature and humidity sensors, pressure sensors, optical sensors and so on.

- **Wireless Technologies:** the most commonly known are Bluetooth, LTE, Wi-Fi.

IoT combines the "brains" of embedded controls, the modularity of smart sensors, and the ubiquitous connectivity of wireless communication.

As a result, the automated manufacturing systems as of today and even more in the future are going to be a composition of software, data, which form a digital twin along with the electronics and mechanical hardware of the physical systems. A full integration of cyber physical systems and availability of a digital twin, i.e., a cyber representation which could be managed in a cyber or physical way, is still a future vision as a number of frontiers of the automation technology as of today need to be overcome.

The following issues have been raised and need attention:

- Improved techniques are required for establishing the communication between the components of the cyber physical systems in order to reduce the effort for interoperability

- New types of standards, e.g., for the semantic processing of information, are required but are difficult to conceive.
- Means to manage the complexity of very large automation systems a yet to be invented.
- The topics of analytics, machine learning and artificial Intelligence need to be deployed to enable Self-X functionalities and present limitation in automatic adjustment of systems.

7.5 Establishing a communication

Different technologies enable wireless communications, the most common are:

- Barcodes, QR codes, Data Matrix etc
- Bluetooth
- Wi-Fi
- GSM, LTE, 4G, 5G etc
- Near Field Communication (NFC)

- Radio-frequency Identification (RFID)

Position tracking can be also considered a way to interconnect systems, as they provide "info" about their position. In this sense, additional technologies are:

- Global Positioning System (GPS), used for outdoor tracking
- Infrared (IR), used for indoor tracking
- Digital Camera

The adoption of the Internet of Things is driven by the falling cost of smart sensors and the ease with which, as they become ever smaller, they can be deployed into new products.

RFID is probably one of the most widely used technology, due to its simplicity and low cost. In 2014, the world RFID market was worth US$8.89 billion, up from US$7.77 billion in 2013 and US$6.96 billion in 2012. This figure includes tags, readers, and software/services for RFID cards, labels, fobs, and all other form factors. The market value is expected to rise to US$18.68 billion by 2026.

7.6 IoT Protocol & Standards

We have already shown how interconnectivity between CPS implies the usage of different technologies (e.g., RFID, GSM, GPS, BLE etc.) each one relying on different protocols and standards. Therefore, it becomes clear how important is to establish the right standard in order that systems can communicate and interact.

Just to make an analogy with human communication, everyone recognizes how useful is to have a common language to speak when we meet people from other countries. This aspect became crucial with globalization. Somehow English has been recognized as a "Lingua Franca", i.e., as "a common means of communication for speakers of different first languages"

With the same logic, it would be useful to have a "Lingua Franca" to enable system to speak the same language.

7.7 Applications of IoT

7.7.1 Industrial Applications

Asset Tracking: it refers to the method of tracking physical assets, either by scanning barcode labels attached to the assets or by using tags using GPS, BLE or RFID which broadcast their location. These technologies can also be used for indoor tracking of persons wearing a tag.

Asset inventory

- Helps to manage physical capitals and allow to make more informed decisions regarding inventory, such as when to repair or replace items;
- Maximizes employee and equipment efficiency;
- Reduces equipment downtime through better planning;
- Prevents theft and enhance security of items.

Reconfigurable Manufacturing Systems: Reconfigurable manufacturing systems (RMSs) are attractive options for handling product personalization, as the system can be continuously reconfigured in accordance with the demanded volumes and products. However, the development of the RMS is a particularly challenging task compared to the development of a traditional manufacturing system[44].

Predictive maintenance: Machines are subjected to periodical maintenance to prevent in service failures which can cause a lot of problems in a production system:

- low productivity
- higher cost
- unpredicted repair time, causing longer lead time and in the end lower customer satisfaction

Total Productive Maintenance (TPM) is a Lean technique and it focuses on keeping all equipment in top working condition

---

[44] Bejlegaard M et al, Reconfigurable Manufacturing Potential in Small and Medium Enterprises with Low Volume and High Variety, 3rd International Conference on Ramp-up Management (ICRM), Procedia CIRP 51 (2016), 32–37

to avoid breakdowns and delays in manufacturing processes. TPM can be considered a Level II approach (Planned).

Today, the combined benefits of the cloud, including lower cost of ownership, high-power computing resources and IoT connectivity capabilities, have enabled more advanced AI-based systems that leverage machine learning to generate higher-value analytics than more basic, model-based systems. Which means, in other words, that it is now possible to predict breakdowns caused machine failures by monitoring the health of key subsystems, such as spindles, electric motors, bearings and so on.

Statistical Process Control: SPC is a widely used Quality tool in Industry to monitor process deviations. The basic idea is to monitor key process variables (KPV) to prevent defects. This mean that data must be recorded and subsequently analyzed. IoT enables automatic solution for Data Collection. A typical example are smart gauges which can automatically collect measured values. The same principle is applied on smart torque wrenches which record and collect torque an angle value. Such data can be then automatically analyzed and eventually real-time feedback (e.g., warning or corrective action) can be provided by AI system.

Mistake proofing: also called Poka-Yoke, mistake (or error) proofing is a technique used by lean manufacturing to prevent mistakes. Some examples are:

- barcode, QR code or RFID instead of human readable methods prevent typos;
- smart torque wrenches use RFIDs to automatically set different torque values for different applications;
- asset tracking can be used to prevent people from taking wrong parts or assemble parts in the wrong position by tracking the arm

7.7.2 Other Applications

Smart home: home automation implies the application of smart technologies in a home environment. Some examples are:

- **Energy control**: it is possible to have remote control of all home energy monitors over the internet incorporating a simple and friendly user interface;
- **Light control system**: a "smart" network that incorporates communication between various lighting system inputs and outputs, using one or more central computing devices;
- **Using Voice control devices** like Amazon Alexa, Google Home or mobile applications to manage coffee machines, ovens, fridge etc.;

Healthcare: medical data can be collected throughout different sensors and devices (e.g., smart bands used for running). Below a couple of interesting applications:

- Alcon (part of Novartis) has licensed Google's smart lens technology which involves non-invasive sensors embedded within contact lenses. The lenses may eventually be able to measure glucose levels of diabetes patients via their tears and then store the information in a mobile device, though Novartis backtracked on a plan to test the system in 2016

- Some dental insurances provide connected toothbrush to offer special deals to their affiliates by monitoring their daily cleaning behavior.

Remote monitoring: big constructions like Dams, Bridges, Buildings, Tunnels and so on can be constantly monitored in real-time.

Shopping: Amazon introduced computer vision and machine learning inside supermarkets, called Amazon GO. You get access through your mobile phone scanning a QR code, then cameras and algorithms create your virtual cart. No final cashier, simply leave the shop and you will be charged automatically.

Chapter 8: Horizontal & Vertical IT Systems Integration

8.1 Introduction 169

8.2 Horizontal IT Systems Integration 173

8.3 Vertical IT Systems Integration 177

8.4 Inter-Organizational IT Systems Integration 178

Chapter Summary

With Industry 4.0, companies, departments and functions will become much more cohesive, as cross-company, universal data-integration networks evolve and enable truly automated value chains. In this chapter, a short glimpse on IT systems and benefits on their integration within the company and the supply chain will be provided.

Keywords: IT systems integration, horizontal integration, vertical integration, inter-organizational integration

8.1 Introduction

Organizations have traditionally been structured around business functions, supported by individual IT systems, using most of the time different applications and technologies, such as programming languages or protocols.

However, today's trends are changing: competitive businesses need high level of integration among all divisions. Therefore, IT support must not be restricted to a single business function, but cut across many different functional areas. These trends require for the integration of current functionality-oriented systems, and this integration needs to be delivered within short time[45]. In the factory of the future, networking and interconnectivity are key components, therefore intra-organizational and inter-organizational cooperation and communication will increase significantly. Workers will collaborate and communicate without borders by using smart

---

[45] Gehrke L. et al, A Discussion of Qualifications and Skills in the Factory of the Future: A German and American Perspective, VDI and ASME, April 2015

devices which connect them in real-time to their co-workers and workplace tools as needed.

Figure 15 shows the 3 main levels of an organization:

- Shopfloor it is the lowest level. Operations, Supply Chain, Quality, Design Manufacturing Engineering are typically involved
- Organization and planning are the intermediate level where organizational activities at all functional levels are performed
- "Standards" is the highest level, where company standards are defined and strategic decisions are taken

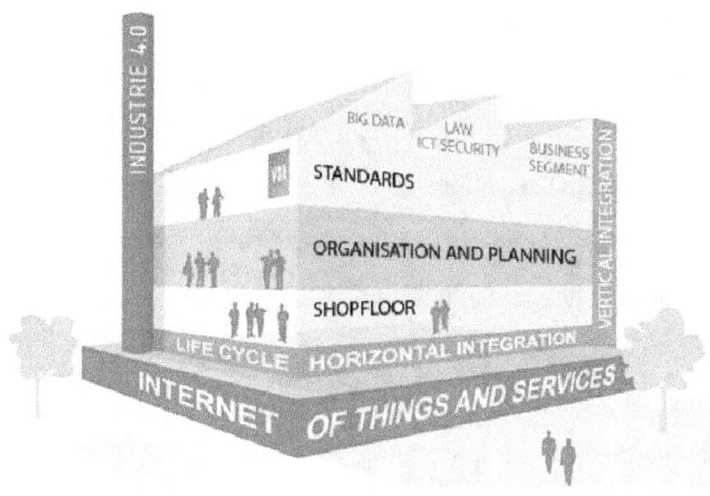

Figure 15: Vertical and Horizontal integration under Industry 4.0[46]

The product portfolio of an organization IT department is becoming wider and wider:

- HR typically uses dedicated applications to manage employees' payslips, career path and similar activities

---

[46] Gehrke L. et al, A Discussion of Qualifications and Skills in the Factory of the Future: A German and American Perspective, VDI and ASME, April 2015

- Design Engineering uses CAD, Simulation and PLM software to design and manage products
- Manufacturing Engineering uses similar software, including CAM to program NC machines and MES to interface with Operations
- Quality Engineering uses PLM and dedicated software to manage Quality documentation
- Operations uses MES systems to manage everyday activities and ERP to plan workloads
- Supply Chain uses applications to manage material and keep relationships with suppliers
- Customer Support has specific applications to manage relationship with customers
- Program Management and Planning use specific software and applications, usually ERP and PLM are shared with other divisions

These are only some examples of applications required by different departments in a mid-large organization. Some software solutions already integrate different functionalities: for example, ERP systems embed functionalities of MES and PLM systems and vice versa.

A high level of integration and data-exchange imply several benefits:

- data becomes easily available at different organization levels
- better data driven decision making
- less management costs
- less time spent in retrieving and collecting data
- less complexity in managing interconnectivity between different systems
- higher efficiency

8.2 Horizontal IT Systems Integration

Horizontal System Integration implies the integration of IT systems at the same organization's level. According to Wangler and Paheerathan *"a typical example of horizontal integration is Supply Chain Management, in which an organization tries to optimize the complete set of activities of order entry, purchasing, production, shipment etc to minimize lead time and costs for production and at the same time maximize value for the customer".[47]*

Another concrete example is the production environment. Operations is the core of manufacturing companies selling products:

- in modern organizations, concurrent design and manufacturing activities are required to make better, faster and cheaper products. CAD and PLM (Product Lifecycle Management) software are typically involved in this process
- once the product has been designed, it is necessary the design and implement the proper method of manufacturing and assembly. CAD, CAM, PLM and MES software are commonly used, as well as ERP to other dedicated tools to plan and buy materials and contact suppliers to perform purchase orders. In parallel, a Quality Plan must be deployed

---

[47] Wangler B., Paheerathan S J, Horizontal and vertical integration of organizational IT systems, Information Systems Engineering, The Pennsylvania State University, 2000

- Parts are then produced following the manufacturing and assembly routes released by MES tools.

A Product Lifecycle Management system (PLM) is a business computing solution that aims to manage the entire life cycle of the product on a single computing platform: from the time it is created and designed, through the prototype phase, first pre-series, manufacturing and maintenance-after-sales service. A PLM system coordinates how people create and use product information in their daily processes.

A Manufacturing Execution System (MES) is a software tool that functions as an extension of the ERP system, but oriented to the planning and execution of production. In this way, while the ERP system determines what has to be manufactured, the MES system provides the necessary functions for the management of key areas in a plant, such as people, materials, processes, quality, traceability and maintenance.

An Enterprise Resource Planning (ERP) is the integrated management of core business processes, often in real-time and mediated by software and technology. ERP provides an integrated and continuously updated view of core business processes using common databases maintained by a database management system. ERP systems track business resources—cash, raw materials, production capacity—and the status of business commitments: orders, purchase orders, and payroll. The applications that make up the system share data across various departments (manufacturing, purchasing, sales, accounting, etc.) that provide the data. ERP facilitates information flow between all business functions and manages connections to outside stakeholders.

It is evident that the integration of aforementioned tools may simplify otherwise complex activities and make the data and information flow smoother. Some organizations think that expensive ERP like PeopleSoft, SAP, Baan, etc. may support better the inter-process chain integration, but this is actually not true for 2 main reasons:

- there is no ERP solution that will provide all the functionality an organization requires
- there is always a tendency to maximize the return of past investments on information systems

8.3 Vertical IT Systems Integration

Vertical System Integration means the integration of IT systems to support the information flow through all administrative levels of an organization. An example is the data flow required for production: *"frequently different operating systems and networking technologies are used. These systems need to be fed with control data stemming from higher level planning and scheduling systems while the lower-level applications need to collect data and pass them upwards"*.[48]

Operations has the primary role to produce the number of parts (throughput) required by the organization with assets and instructions provided by Manufacturing Engineering Department. The final outcome can be:

- keep asset portfolio and the number of workers (considering specific skills as well) as-is

---

[48] Wangler B., Paheerathan S J, Horizontal and vertical integration of organizational IT systems, Information Systems Engineering, The Pennsylvania State University, 2000

- improve capabilities by investing money in R&D and new machines
- improve capacity by purchasing new state of the art machines

This type of analysis is normally performed at the intermediate level, while decisions are taken at the top level. Therefore, it is evident how important is to guarantee the proper data flow in both directions (top-down and bottom-up). Typical tools used to manage data flow throughout all different organization's levels are ERPs, MESs and PLMs.

8.4 Inter-Organizational IT Systems Integration

Enterprises need to exchange information in order to collaborate and negotiate. For example, an organization needs to contact suppliers to provide the required assets (e.g. tools, fixtures, machines and so on) against a purchase order from a customer. *"Supply chain logistics and fulfillment companies integrating with their customers' fulfillment and shipping systems, or financial services firms integrating with retailers of financial products are just few more examples of this multi-enterprise integration".[49]*

Since different companies use different tools, systems and platforms, the key issue in inter-organizational integration is to get the data of one application of an organization matching with another application of another organization. There are 2 main ways to achieve this purpose:

- use the same tool or platform. Some examples are:
  - in the supply chain is the use of Electronic Kanban Systems
  - use file/data sharing platforms

- use same data standards and protocols:
  - use same CAD formats
  - use same data protocols. For example, in the gear industry, Gear Data Exchange (GDE) is becoming a sensitive topic

---

[49] Wangler B., Paheerathan S J, Horizontal and vertical integration of organizational IT systems, Information Systems Engineering, The Pennsylvania State University, 2000

Chapter 9: Simulation

9.1 Introduction 183

9.2 Discrete Event Simulation 185

9.3 Process Simulation 192

Chapter Summary

Simulation tools are fundamental in engineering to predict the behaviour of new products, therefore reducing the associated risks before entering into service. However, if simulation is widely used in Design Departments, however we can't say the same in Manufacturing. In this chapter we will introduce some of the most important simulation tools that are now available to implement new facilities or to optimize existing production lines. In this sense, Discrete Event Simulation is a fundamental tool.

Keywords: Simulation, Discrete Event Simulation, Process Simulation

9.1 Introduction

"Physical testing costs at least five to six times the cost of product development resources on vehicle projects. The only way to meaningfully reduce the cost of physical testing is with simulation."

Dominic Gallello, president of MSC Software.

"The digital manufacturing building is based on simulation processes for testing new ideas and options before actual implementation of these ideas. With simulation models, we can explicitly visualize how an existing operation might perform under varied inputs and how a new or proposed operation might behave under same or different inputs, analyze the material flow and optimize plant lay-out. Today simulation can be used for decision support with supply chain management, workflow and throughput analysis, facility layout design, resource usage and allocation, resource management and process change"[50]

---

[50] Kokareva V.V. et al, Production Processes Management by Simulation in Tecnomatix Plant Simulation, Applied Mechanics and Materials Vol 756 (2015) pp 604-609

Simulations tools have been traditionally used in the engineering world since decades. Some examples are:

- CAD models, which are normally used as digital mock up to simulate geometries and encumbrances
- CAM systems, which are used to simulate machining processes, typically the tooling path
- Multiphysics simulations, which are usually numerical implemented with discretization methods such Finite Element Method, Finite Difference Method, and Finite Volume Method. Many software packages mainly rely on the finite element method or similar commonplace numerical methods for simulating coupled physics: thermal stress, electro- and acoustic- magneto mechanical interaction

9.2 Discrete Event Simulation

A Discrete-Event Simulation (DES) models the operation of a system as a discrete sequence of events in time. Each event occurs at a particular instant in time and marks a change of state in the system. Between consecutive events, no change in the system is assumed to occur; thus, the simulation can directly jump in time from one event to the next. DES is a popular decision support tool, often used to enhance understanding of interactions within the simulated system, and identifying system issues (e.g., bottlenecks and long lead times) and solutions. A DES study starts with the building of a virtual model of a physical system, current or future, and then simulating and analyzing the results of the virtual model to gain insights into the physical system. DES is commonly used for the simulation of dynamic systems (e.g., manufacturing and supply chains).

A Discrete Event Simulation is an effective approach for a confident decision making based on data and experiments. Simulation is able to answer critical questions to different levels:

- **Plant designer**: how to satisfy the expected performance of the system for the client?
- **Plant manager**: how to adapt the production system to new products & improve it?
- **Production manager**: how to optimize the production plan? Impacts of the changes in the line?

A DES can also provide important support in different phases in manufacturing. In Table 7 you can see what type of support a DES can provide during the design phase, reconfiguration phase and production planning phase.

	Design	Reconfiguration	Production Planning
Main aspects	Capacity	Reconfig. Costs	Operational Costs
	Capex	Utilization	Inventory
	Flexibility	Reconfig. Time	Deadlines
	Utilization		Reliability
Simulation Model	High Fidelity Analysis	Existing conditions	Production Status
	Uncertainty	Systems evolutions	Customer orders
	Different designs	Different designs	Shopfloor conditions

Table 7: How DES can support different phases in manufacturing

In the design phase, some of the main aspects to take under considerations are the production capacity, the capital expenditure (capex), the flexibility that a production system should guarantee and utilization of machines and workers. In this phase, a DES is able to provide high fidelity analysis of different designs, substantially reducing the associated uncertainty.

The reconfiguration phase occurs when changes must be introduced in our production system, due to the introduction of new products or to an increase in capacity or to optimize the system. In this phase, DES is an important tool to model existing conditions and evolutions, as well as to explore different options.

In the production planning phase, a DES is able to calculate with extreme accuracy operational costs, verify if deadlines can be met, therefore verify how reliable the production system is to fulfill customer expectations.

In Figure 16: Block scheme of a Discrete Event Simulation process steps a methodology to conduct a DES study is proposed. These 14 stages provide the foundation to construct an informative and validated DES model, while ensuring the model meets the customer needs.

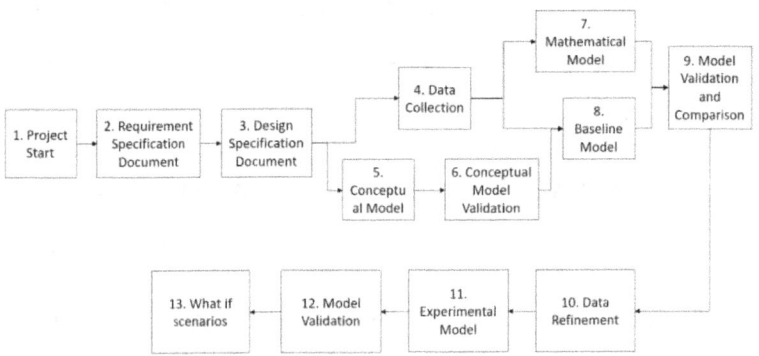

Figure 16: Block scheme of a Discrete Event Simulation process steps

A short description of each stage is provided below:

1. Project start: initial meetings to discuss requirements and scope of study and gain understanding of current process

2. Requirement specification document: capturing exhaustive and prerequisite details to understand and verify project requirements

3. Design specification document: specifying the logic and behaviour of simulation model and verifying the customer

4. Data collection: collecting of what data is available, both numerical and behavioral, which is needed for construction of the model

5. Conceptual model: defining of process flow diagram which shows how the model will be structured and how it will behave

6. Conceptual model validation: validating the conceptual model with the customer to ensure that all the data and processes are captured before model development can start

7. Mathematical model: creating of a mathematical model based on the collected data. This includes outputs such as expected processing times and number of reworks and is used as an internal validation tool for the DES model, checking model results against expectations from the raw data

8. Baseline model creation: creating of a baseline DES model to reflect the current state of the facility

9. Model validation / Test plan: validating and verifying the model using the mathematical model and with input from the customer

10. Data refinement: updating the data and behavior in the model according to new data or time and motion studies that have been performed

11. Experimental model: using the updated data, and customer feedback to update the DES model into a form suitable to explore scenarios;

12. Model validation: validating experimental DES model with the customer. Completion of this stage implies that the model is frozen, and no data will change;

13. "What-if" scenarios: using the experimental DES model to produce results for scenarios of interest that have been agreed upon by the customer

According to Kokareva et al.[51], main benefits for the production planner include:

- enhance productivity of existing production facilities by as much as 20%
- reduce investment in planning new production facilities as much as 30%

---

[51] Kokareva V.V. et al, Production Processes Management by Simulation in Tecnomatix Plant Simulation, Applied Mechanics and Materials Vol 756 (2015) pp 604-609

- cut inventory and throughput time by as much as 40%
- optimize system dimensions, including buffer sizes
- reduce investment risks through early proof of concept
- maximize use of manufacturing resources
- improve line design and schedule

Different tools are available on the market, but the most popular are:

- Tecnomatix Plant Simulation by Siemens
- Witness by Lanner
- DELMIA by Dassault Systeme
- Simul8 by Simul8 Corporation
- Flexsim by FlexSim Software Products, Inc

9.3 Process Simulation

In addition to Discrete Event Simulations, the market offers several tools to simulate in very detail each step of a process. At the beginning of this chapter, we have already mentioned some common tools used in manufacturing like CAM simulations or Multiphysics simulations.

Another type of simulation which is also recommended to analyze the overall process is the Process Simulation. One of the most common digital tools in this sense is called Process Simulate by Siemens Process Simulate is a digital manufacturing solution for manufacturing process verification in a 3D environment. Main features include:

- Static and dynamic collision detection
- Sequencing of operations
- Assembly and robotic path planning
- Line and workstations design
- Human tasks simulation, like reach envelopes, vision window, postures, ergonomics analysis
- Robot-related simulation, like reach test, process simulation, programming, logic editing and validation
- Virtual commissioning

Main benefits of Process Simulation are:

- Reduce cost of change with early detection and communication of product design issues

- Reduce number of physical prototypes with upfront virtual validation
- Optimize cycle times through simulation
- Ensure ergonomically safe processes
- Reduce cost by re-using standard tools and facilities
- Minimize production risk by simulating several manufacturing scenarios
- Early validation of the mechanical and electrical integrated production processes (PLC and robotics)
- Increase process quality by emulating realistic processes throughout the process lifecycle

Chapter 10: Virtual Reality

10.1 Brief History 197

10.2 How VR works, hardware and software 201

10.3 Main benefits, limitations and applications of VR 208

Chapter Summary

Virtual Reality (VR) is an interactive computer-generated experience taking place within a simulated environment. It incorporates mainly auditory and visual feedback, but may also allow other types of sensory feedback like haptic. This immersive environment can be similar to the real world or it can be fantastical. In this chapter, we will see how this technology evolved through years, what the two main VR technologies are and how they work. In the final section, we will present some benefits, challenges and real applications in industrial contexts.

Keywords: Virtual Reality, VR, Headset Technology, CAVE

10.1 Brief History

- **1838:** the stereoscope is invented by Sir Charles Wheatstone. This is our first foray into 3D graphics

- **1950:** Morton Heilig's Sensorama. In the mid-1950s cinematographer Morton Heilig developed the Sensorama (patented 1962) which was an arcade-style theatre cabinet that would stimulate all the senses, not just sight and sound. It featured stereo speakers, a stereoscopic 3D display, fans, smell generators and a vibrating chair

- **1960:** the first VR Head Mounted Display. Morton Heilig's next invention was the Telesphere Mask and was the first example of a head-mounted display (HMD)

- **1961:** two Philco Corporation engineers (Comeau & Bryan) developed the first precursor to the HMD as we know it today – the Headsight. It incorporated a video screen for each eye and a magnetic motion tracking system, which was linked to a closed-circuit camera

- **1965:** The Ultimate display by Ivan Sutherland. Ivan Sutherland described the "Ultimate Display" concept that could simulate reality to the point where one could not tell the difference from actual reality. His concept included: 1) a virtual world viewed through an HMD and appeared realistic through augmented 3D sound and tactile feedback; 2) computer hardware to create the virtual world and maintain it in real time; 3) the ability users to interact with objects in the virtual world in a realistic way

- **1968:** The Sword of Damocles is invented by Ivan Sutherland. His invention tracked user head motion and overlayed vision with synthetic computer-generated graphics of primitive 3D object wireframes

- **1987:** Jaron Lanier, founder of the visual programming lab (VPL), coined the term "virtual reality". Jaron developed a range of virtual reality gear including the Dataglove (along with Tom Zimmerman) and the EyePhone head mounted display. They were the first company to sell Virtual Reality goggles (EyePhone 1 $9400; EyePhone HRX $49,000) and gloves ($9000)

- **1991:** Virtuality Group Arcade Machines: The Virtuality Group launched a range of arcade games and machines. Players would wear a set of VR goggles and play on gaming machines with real-time (less than 50ms latency) immersive stereoscopic 3D visuals. Some units were also networked together for a multi-player gaming experience
- **1995:** Nintendo Virtual Boy (originally known as VR-32): it was a 3D gaming console that was hyped to be the first ever portable console that could display true 3D graphics. It was first released in Japan and North America at a price of $ 180 but it was a commercial failure despite price drops. The reported reasons for this failure were a lack of color in graphics (games were in red and black), there was a lack of software support and it was difficult to use the console in a comfortable position. The following year they discontinued its production and sale.

- **2012:** Oculus Rift Kickstarter is launched and backed. Oculus would later be bought by Facebook for $2 billion

- **2014:** Google Cardboard released.

- **2015:** The HTC Vive was unveiled during HTC's Mobile World Congress keynote in March 2015. Development kits were sent out in August and September 2015, and the first Consumer version of the device was released on April 5th, 2016.

- **2016:** Google Daydream was announced.

- **2018:** At the Facebook F8 Developer Conference, Oculus revealed the Half Dome – a headset with a 140-degree field of vision

10.2 How VR works, hardware and software

There are basically 2 types of Virtual Reality technologies:

- Cave Automatic Virtual Environment (CAVE)
- Headset Technologies

CAVE: The Cave Automatic Virtual Environment (CAVE) is a video theater situated within a larger room. The walls of a CAVE can be made up of

- rear-projection screens, or
- flat panel displays

The projection systems are very high-resolution due to the near distance viewing which requires very small pixel sizes to retain the illusion of reality. The user wears 3D glasses inside the CAVE to see 3D graphics generated by the CAVE. People using the CAVE can see objects floating in the air, and can walk around them, getting a proper view of what they would look like in reality. This is made possible by the use of infrared cameras. CAVE user's movements are tracked by the sensors typically attached to the 3D glasses and the video continually adjusts to retain the viewers perspective. Computers control both this aspect of the CAVE and the audio aspect. There are typically multiple speakers placed at multiple angles in the CAVE, providing 3D sound to complement the 3D video.

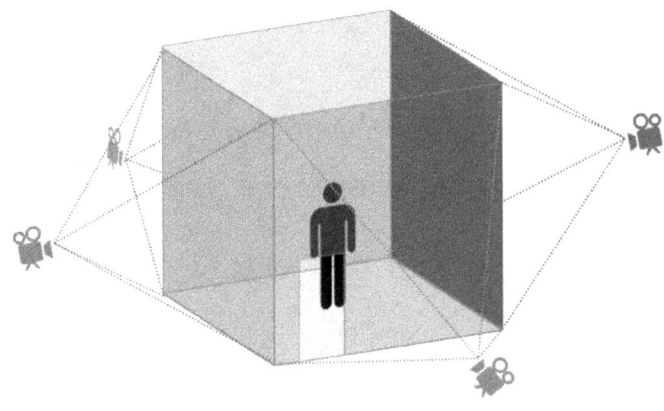

Figure 17: Pictorial representation of CAVE technology[52]

Headset Technologies: Headset Technologies are more familiar to the big public as it provides a cheaper solution to experience VR. Headsets comprise a stereoscopic head-mounted display (providing separate images for each eye),

---

[52] https://www.themarketingtechnologist.co/virtual-reality-connecting-unity-to-the-cave/

stereo sound, and head motion tracking sensors (which may include gyroscopes, accelerometers, structured light systems, etc.) Base on their tracking method, Headset technologies can be divided in 2 main categories:

- **3 DoF:** Tracking will track how the headset is rotated, so it will know where we're looking. Most mobile VR platforms use 3DoF tracking. Inertial Measurement Units (IMUs) are used to track the rotation. IMUs use a gyroscope, accelerometer and a magnetometer to detect rotations. All modern 3DoF tracking uses an IMU.
- **6 DoF:** It will track rotation PLUS position. It will know where we're looking and where we are in the world. This is how high-end VR systems, like the HTC Vive and Oculus Rift, allow you to move around the room. There are many different ways to accomplish 6DoF tracking. The Oculus Rift uses a large number of LEDs and external cameras that build a model and then it tries to fit a 3D model of the headset with the 2D model it sees of the camera. Then it uses information from an internal IMU to calculate head position. The HTC Vive uses infrared lasers to measure the time it takes to sweep vertically and horizontally across a

photo sensor. It uses the data it gets and an internal IMU to determine where your head and hands are located.

Another classification of VR Headset-based technologies is:

- **Mobile VR:** Mobile VR uses IMU tracking, so it only has 3DoF tracking. It can only track your head's rotation – not it's position. It's also "untethered", meaning that there's no wires connecting your headset to a computer. You can use it anywhere. It has less powerful graphics and is powered by a battery.
- **Desktop VR**: Desktop VR uses 6DoF tracking and is typically tethered. It's plugged into a computer and uses the computer to do the graphics work and to get its power. The graphics are generally more impressive. Most desktop VR headsets also come with hand controllers that allow for more interactivity and a more immersive experience.

A very popular VR set is provided by HTC, the HTC Vive, which are already widely used in big companies for different purposes.

The HTC Vive starter kit, for example, is composed by the following hardware:

1 Headset:

- Screen: Dual AMOLED 3.6" diagonal
- Field of view: 110 degrees
- Sensors: SteamVR Tracking, G-sensor, gyroscope, proximity

2 Controllers:

- Sensors: SteamVR Tracking
- Input: Multifunction trackpad, Grip buttons, dual-stage trigger, System button, Menu button

2 Base stations

- IR laser grid emission
- 360-degree play area tracking coverage

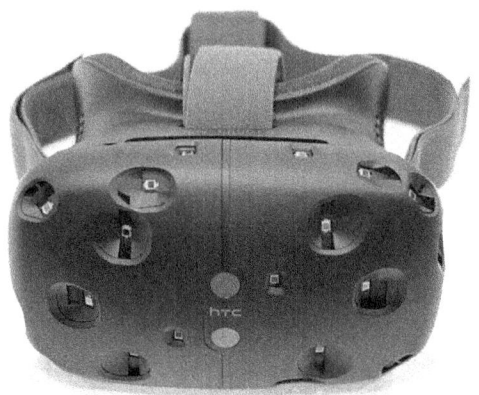

Figure 18: HTC Vive Headset

However, there is another super cheap way to experience VR with mobile VR based technologies. Indeed, you can buy Google Cardboard for about 10 $ on Amazon, for example. Then, what you have to do is to put your smart mobile inside the Cardboard and launch a VR experience. It's plenty of App you can download for free, but also you can watch Youtube videos in VR!

Headset-based VR Development: generally, developing for VR requires writing some code, using an editor, or a combination of both. An editor, or game engine, is a software development framework that makes it easier to develop games. Game engines typically provide 3D rendering, physics, sound, scripting, animation, asset management, and more. This will enable us to develop quickly and learn. Other popular game engines are:

- Unreal Engine
- Cryengine
- Lumberyard

You can also use native development tools like OpenGL and Microsoft DirectX. These options offer more flexibility, but typically have much longer development times and writing more code. WebVR is also an option. However, it's not yet good enough for production apps.

CAVE VS Headset Technologies

In my professional career I had the opportunity to try both solutions. My preferred solution is the Headset Technology for different reasons:

- it is much cheaper compared to a CAVE implementation

- the virtual experience is different: the CAVE provides probably a more realistic experience, while the Headset is like being in a videogame. From my point of view is difficult to say which is better. Personally, I prefer the Headset solution
- Portability. A CAVE is fixed, while the headset is portable

10.3 Main benefits, limitations and applications of VR

Main Benefits

- Risk mitigation assessment in an immersive environment for an experience close to real life.
- Safe environment to experience situations without risks.
- Low-cost technology, compared to physical training.
- Gamification helps in the learning process.

Main Limitations

- Gravity can´t be experienced: during a training session, for example, the gravity effect can play a major role (e.g., lifting objects).
- People are not familiar with this technology; therefore, they are reluctant in investing even a minimum budget for VR.

Main Applications

- **Training**: as already mentioned, this is probably the main application for VR. Aerospace companies are already using this solution for MRO training.
- **Factory planning**: a VR simulation provides an engaging and low risk environment to test solutions and scenarios before capital investments.
- **Product Development Assessments**: VR provides a more realistic environment rather than assessing 3D models on a computer.
- **Marketing**: more and more companies are offering VR to advertise their products in trade fairs

Chapter 11: Other Technologies

11.1 Smart Human Machine Interface 213

11.2 The Digital Twin 220

11.3 Blockchain 225

Chapter Summary

In the previous chapters, we have provided a full description of the key 4.0 technologies identified by the Boston Consulting Group. However, other smart technologies are available on the market and some of them are becoming more and more popular in different industrial sectors.

In this chapter, we will describe 3 additional technologies or trends that are worth mentioning ad describing more accurately: Smart Human Machine Interface, Digital Twin and Blockchain.

Keywords: Human Machine Interface, Wearables, Digital Twin, Blockchain

11.1 Smart Human Machine Interface

A Human Machine Interface (HMI) is basically a device which interfaces the machine and the operator. The main 4 functions are:

- allow the machine/system to display its status to the operator
- allow the machine/system to display instructions to the operator
- allow the machine/system to display outcomes
- allow the operator to input actions to the machine/system

Although touch screens are widely used especially in machining industry, several different devices can be now used as well. Most of them are part of the "Wearables" category. Typical wearables are glasses, watches/smartphone and gloves.

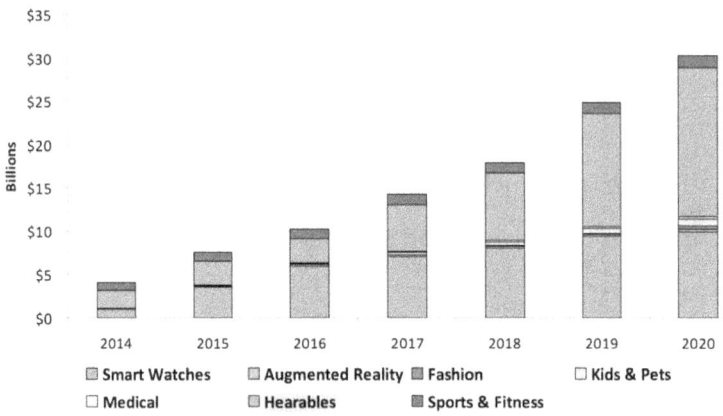

Figure 19: Global Revenue from Smart Wearables and Hearables[53]

Main benefits of wearables are:

- **Hands-free:** workers don't need special devices to handle, for example a scanner gun to scan barcodes, or a monitor to check work instructions or input data. This will increase productivity

---

[53] https://commons.wikimedia.org/wiki/File:Global_Revenue_from_Smart_Wearables_and_hearables.png

- **Safety:** safety information can be displayed or in general communicated to the operator in real time in case of fire or similar events
- **Confirmation technology:** selection of work step confirmation technologies that you need
- **Image recognition:** identification of objects just by the power of sensors
- **Localization and navigation:** navigation and positioning for your frontline worker to stay on track. This might be a problem and it is recommended to agree the utilization of this options with the employer and unions
- **Remote assistance:** a camera embedded on smart glasses allows technicians to provide instructions remotely
- **Live video call:** communication between workers and technical assistance whenever needed
- **Documentation:** transformation of work activities into instant documentation with zero touch
- **Ergonomics:** exoskeletons may reduce significantly health problems due to ergonomic issues
- **Reporting & Analytics:** monitoring of all work being executed by your frontline workers with simple tools

Tablets: Tablets are widely used by people in their everyday life. They combine features of a laptop and portability. Nowadays 2-in-1 laptop may provide basically the same benefits. Tablets can be used by quality supervisors to daily monitor the WIP status, document non conformances, and to connect to machines to monitor their "health". Emails can be immediately sent in case of issues. Special tablets are available for industrial utilization: these tablets are built to survive to specific environment or products (e.g., corrosion protection oil, lubricant, coolant etc.)

Smart Glasses: smart glasses are wearable computer glasses that add information alongside or to what the wearer sees. Like other computers, smart glasses may collect information from internal or external sensors. It may control or retrieve data from other instruments or computers. It may support wireless technologies like Bluetooth, Wi-Fi, and GPS. A smaller number of models runs a mobile operating system and function as portable media players to send audio and video files to the user via a Bluetooth or Wi-Fi headset. Some smart glasses models, also feature full lifelogging and activity tracker capability. Some applications are:

- smart glasses can be used to identify the asset position and scan barcode/ QR code in a warehouse
- more complex solutions can be adopted to display work instructions and guide the operator
- complex smart glasses solutions may provide support for training and designing exploiting a high level of augmented reality

Smartphones and smartwatches: smartphones and smartwatches are small and powerful computers. Attached to an arm or a wrist, are often used by runners to track performances and listen to the music. However, they can be used in a shopfloor environment for all different purposes listed above. Moreover, special apps can be used for additional activities, possibilities are limitless.

Wristbands: Amazon has patented designs for a wristband that can precisely track where warehouse employees are placing their hands and use vibrations to nudge them in a different direction. When someone orders a product from Amazon, the details are transmitted to the handheld computers that all warehouse staff carry. Upon receiving the order details, the worker must rush to retrieve the product from one of many inventory bins on shelves, pack it into a delivery box and move on to the next assignment[54]. Similar solutions are used by the German company Sarissa that developed a Local Positioning System (LPS) which combines advanced ultrasound technology, easy-to-use software and a powerful, open interface architecture. It can be used as a mistake proofing solution to track, for example, if the worker is picking up the right components from a kitting rack.

Exoskeletons: we mention exoskeleton here although it would be more correct to locate them in the Human – Machine – Collaboration category. Anyway, exoskeletons are armor that serve to help operators to maintain prolonged postures and support weight in a more ergonomic way. Some exoskeletons available on the market are:

- **SuitX**, a spinoff of the University of California, develops the so-called modular exoskeleton MAX, consisting of three parts for areas of common injuries in the workplace:

---

[54] https://www.theguardian.com/technology/2018/jan/31/amazon-warehouse-wristband-tracking

Shoulders, Lumbar and Knee. The launch cost is around $3000, varying depending on the units purchased

- **LegX**, a structure that goes from the hip to the feet with regulation of the degree of inclination and that allows the operator, with semi-flexed legs, to reduce the effort to stay standing (the sensation is similar to sitting on a chair) lower back
- **BackX** reduces the weight of the objects that are lifted from the ground by 13 kg
- **ShoulderX** allows you to reduce the effort when supporting weights on the head
- **ExoArm**, developed by two young Slovenian engineers, is an open-source arm at a cost of only 100 €. It has an Arduino heart, and they're trying to program everything with an easy code to understand and modify. They have a first functional prototype capable of lifting weights of 10 kilograms, and the next step will be to improve the design to finally launch it on the market

11.2 The Digital Twin

Digital Twin refers to a digital replica of physical assets, processes, systems and devices. It integrates artificial intelligence, machine learning and software analytics with spatial network graphs to create living digital simulation models that update and change as their physical counterparts' change. A digital twin continuously learns and updates itself from multiple sources to represent its near real-time status, working condition or position. This learning system, learns from itself, using sensor data that conveys various aspects of its operating condition.

11.2.1 Brief History

- **1970**: many authors reported that the concept of a digital twin was first applied during the Apollo 13 program, where engineers on the ground needed to be able to rapidly account for changes to their vehicle while exposed to the extreme conditions in space

- **2003:** the concept of a virtual, digital equivalent to a physical product or the Digital Twin was introduced in 2003 at the University of Michigan Executive Course on Product Lifecycle Management (PLM)[55].

- **2011**: NASA and the US Air Force published two papers on digital twins. They discussed the concept of a digital twin on a structural level to help predict fleet maintenance. These papers are two of the most highly cited documents on the topic, and are recognized as being the first time the phrase was taken seriously by both industry and academia. Subsequently the use of the term declined

- **2016**: the term started spread among the industry community. More recently, the term has seen a marked increase in search activity, most likely due to the general adoption of the term by industry and marketing teams

---

[55] Grieves M, Digital Twin: Manufacturing Excellence through Virtual Factory Replication, March 2015

11.2.2 Components of a Digital Twin

According to a report by the High Value Manufacturing Catapult Visualisation and VR Forum[56], we can identify the following components of a Digital Twin:

Required:

- **A Model:** a model of the physical object or system, which provides context. The provision of a 3D model is not a requirement for the creation of a digital twin. In some cases, it can add some value, but this value is derived from the capability to derive a greater understanding of the contextualisation of the data presented. However, this is not always the case, and in some cases, a 3D model will be excessive to requirements.
- **Connectivity:** connectivity between digital and physical assets, which transmits data in at least one direction

---

[56]https://www.amrc.co.uk/files/document/219/1536919984_HVM_CATAPULT_DIGITAL_TWIN_DL.pdf

- **Real-time:** The ability to monitor the physical system in real-time.

Optional:

- **Analytics:** the optional logic for a digital twin may (and often will) include rule engines or complex-event processing that are applied to incoming IoT data. These logic elements may generate alerts or triggers that orchestrate workflows and various forms of descriptive analytics to identify when thresholds are exceeded; they can also drive predictive analytics that provide inputs to enterprise stakeholders.
- **Control:** not all digital twins will have the ability to control an object but, when they do, they will connect via the object's specific control system. On-board actuators, electronic switches and other digital-to-analog physical devices make up the control systems
- **Simulation:** a sufficiently detailed digital twin may be used by an enterprise to model the current and future behaviour of an object in a variety of conditions and configurations, anticipate failure and optimal operation

modes, or identify optimum schedules for operation, refueling or maintenance.

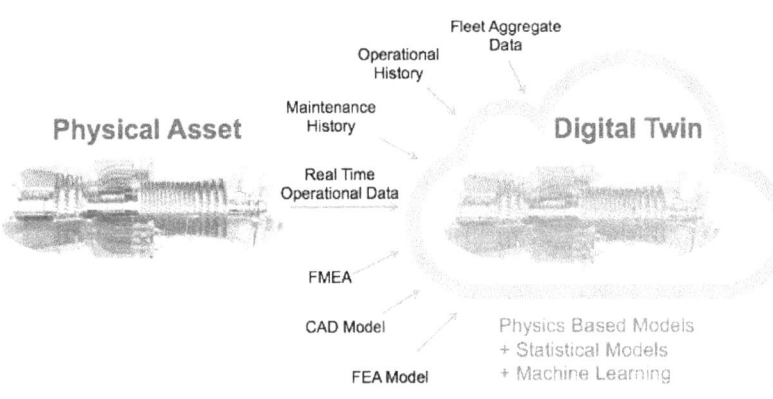

Figure 20: concept representation of a Digital Twin[57]

11.2.3 Some Applications

- **Remote process monitoring**: through the development of digital twin systems, processes can be monitored remotely

---

[57] https://www.bangkokbankinnohub.com/digitaltwin/

- **Remote process controls**: the enablement of control capability from within the digital twin system for remote viewers
- **Predictive analysis**: through the aggregation of historical data, combined with the real time data feed, it is possible to simulate the future states of production systems, including the development of predictive maintenance models. This enables a great productivity saving through the ability to respond to demand signals rather than either
- **Rapid New Product Development**: through increased customer intimacy greater insight into product or process performance can be developed. This can then be used to influence the next generation developments in physical products or processes.

11.3 Blockchain

A blockchain is a decentralized database; however, this simple definition would leave many people thinking "So what? All that hype for a new type of database?"

Blockchain is not an easy concept to explain, and this is out of scope of this book to enter into too many technical details. In the first chapter of his book "Blockchain: Ultimate guide to understanding blockchain, bitcoin, cryptocurrencies, smart contracts and the future of money ", Mark Gates tries to explain the reason why blockchain is not just a database:

"The common theme from everyday transactions is that we trust the institutions and the centralized databases they maintain to accurately keep a record of our lives. [...] What alternative did people have other than deposit money in what they believed were trustworthy banks and companies? [...] A decentralized database built on the blockchain removes the need for centralized institutions and databases. Everyone on the blockchain can view and validate transactions creating transparency and trust. Trust lays at the core of the blockchain; it provides a system of trust between people without the need for an intermediary involved in the transactions. The

blockchain allows people to transact between each other with anything of value" [58]

11.3.1 Brief History

- **1984**: David Chaum introdeuces the "blind Signature" to guarantee the full privacy of its users. A blind signature is a form of digital signature in which the content of a message is disguised (blinded) before it is signed

- **1990**: David founded DigiCash to create a digital currency

- **1997**: Adam Back proposed Hashcash, is a proof-of-work system used to limit email spam and denial-of-service attacks

---

[58] Gates, Mark. Blockchain: Ultimate guide to understanding blockchain, bitcoin, cryptocurrencies, smart contracts and the future of money. Kindle Edition.

- **1998**: Wei Dai published another paper titled "B-Money, An Anonymous, Distributed Electronic Cash System." The paper outlined the foundations for cryptocurrencies

- **2005**: Nick Szabo proposed Bit Gold, and the idea of Smart Contracts

- **2009**: Bitcoin became more than just an idea in an academic paper when Satoshi Nakamoto created the Bitcoin network along with the first blockchain

11.3.2 How blockchain works

In simple words, in a blockchain-based system, transactions between individuals are associated with a crypted code. In order for a transaction to happen, this code must be decrypted by the miners, who are people in the network who decrypt these codes. The first miner who decrypt the code receives a reward for his effort / work in the form of cryptocurrency (e.g., Bitcoin). This mechanism is called "proof-of-work". The decrypted transaction is then sent to the entire blockchain network to validate the solution and to be recorded in every personal public ledger. When other transactions are decrypted, they are "added" on top of the previous ones, therefore creating a sort of chain (here why the name blockchain). If someone wanted to change an old transaction (block), he should ask to at least 51% of the network to change not only the transaction, but also all following transactions, as they are all linked together. It is evident how difficult this is to make it happen, especially when the network is made of thousands or millions of users. However, this is still considered a risk.

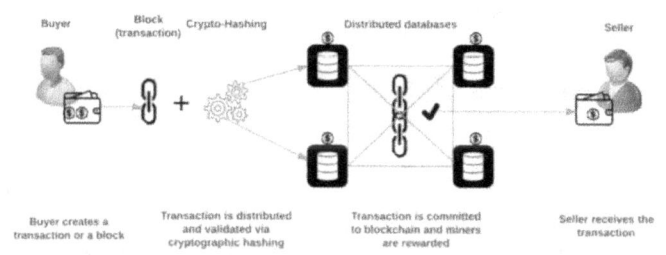

Figure 21: How blockchain works[59]

---

[59] https://commons.wikimedia.org/wiki/File:Blockchain-Process.png

To summarize the process of how a blockchain-based system works (Figure 21):

- A buyer creates a transaction or a block
- To each transaction (or block), a crypto-hashing is associated
- Transaction is distributed
- Transaction is decrypted by miners
- Miners are rewarded
- The decrypted transaction is recorded into public ledgers
- Seller receives the transaction

11.3.3 Smart Contracts[60]

Smart contracts are contracts that are written in computer code and operate on a blockchain or distributed ledger. Smart contracts can be used to exchange anything of value and many

---

[60] Gates, Mark. Blockchain: Ultimate guide to understanding blockchain, bitcoin, cryptocurrencies, smart contracts and the future of money. Kindle Edition.

of the industries utilizing blockchain technology will be using smart contracts. When a smart contract is run on the blockchain, it operates automatically. If the conditions of a contract are met, payments or value are exchanged based on the terms of the contract. Likewise, if conditions in the contract are not met, payments may be withheld if written into the smart contract. Smart Contracts are considered to be the most powerful application of blockchain-based systems for both financial and non-financial applications.

11.3.4 IOTA[61]

IOTA is a cryptocurrency specifically designed for the Internet of things (IoT). Unlike the most popular tokens which are based on proper blockchain architectures, IOTA is based on Tangle: as well as the classic blockchain, Tangle is a distributed network, meaning there is no central instance controlling the currency. Transactions need to create a consensus in order to be validated. In the blockchain one block following the other

---

[61] Kacperczyk, Marcin; Neuefeind, Marvin. Cryptocurrency - A Trader's Handbook: A Complete Guide on How to Trade Bitcoin and Altcoins. Kindle Edition.

in a specific order. Tangle, on the other hand, is entirely different. In the blockchain technology there are different possibilities of mining new blocks in order to create a consensus. On the other hand, if someone wants to make a transaction within the Tangle network, this is not necessary. To be able to submit a transaction, you need to approve two other transactions which need to be validated. Some blocks may be confirmed more than one time, as it is based on a random selection. This means that we do not have the distinction between a node, full node and miner. Every node becomes a miner if he wants to make a transaction. Based on this, it allows the user to make 0-transaction-fee transactions. Another benefit is in contrast to the classic blockchain algorithm is that the more people are using a Cryptocurrency like IOTA (Tangle based), the faster the transaction will be transmitted.

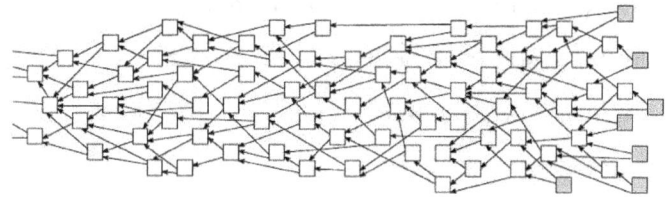

Figure 22: Organization of Tangle blocks[62]

11.3.5 Benefits and challenges

Said that, we can identify some benefits in this process:

- **No intermediaries** (e.g., banks) are required: allowing transactions to occur directly between people instead of involving a third party
- **Transparency**: blockchain provides transparency to all people on the network, with transactions visible to all connected computers
- **Security**: data entered onto a blockchain is immutable, meaning it can't be altered or changed, unless a 51% attack is successful

---

[62] https://commons.wikimedia.org/wiki/File:Tangleimage.jpg

- **Reduced costs**: blockchain could significantly reduce costs in many industries by removing intermediaries
- **Increased transaction speed**: for the same reason, transactions are faster

However, there are also some disadvantages:

- **Energy consumptions**: Professor John Quiggin from the University of Queensland has calculated that every half an hour the Bitcoin network uses the same amount of electricity as the average US household does in an entire year[63]
- **Lack of privacy**: not only is the information not private, but it is also readily accessible at any given moment to anyone using the system
- **Security concerns**: more security can sometimes result in a system being less secure. There are countless examples with cryptocurrencies where someone has forgotten their private key and can't access their money

---

[63] Gates, Mark. Blockchain: Ultimate guide to understanding blockchain, bitcoin, cryptocurrencies, smart contracts and the future of money. Kindle Edition.

- **Risk of 51% attack**: if someone were able to control over 50% of the computers on a blockchain network, they would control the transactions on the blockchain
- **Lack of scalability**: at the current rate of energy consumption, the electricity costs of running a blockchain make it unfeasible to handle the number of transactions by credit card companies

11.3.6 Applications

The most common application of blockchain is the creation of cryptocurrencies and financial transactions. However, the development of the so-called Smart Contracts is opening a wide range of opportunities for non-financial applications as well. Some examples are[64]:

- **Identity management and digital identities**: blockchain-based identification systems provide digital signatures using cryptography

---

[64] Gates, Mark. Blockchain: Ultimate guide to understanding blockchain, bitcoin, cryptocurrencies, smart contracts and the future of money. Kindle Edition.

- **Healthcare and Medical Records**: storing medical records on a shared database would mean that doctors, hospitals, surgeons, nurses, and health professionals would have access to shared data about a patient at any time, saving time and assisting them to make more comprehensive decisions when treating a patient
- **Academic Certificates:** the blockchain would create transparency around students' academic records and qualifications
- **Cloud Storage**: cloud storage currently requires a lot of trust in third-party companies. Centralized cloud storage systems are vulnerable to attack and passwords can easily be obtained through basic hacking or scamming methods
- **Property & Rental records**: blockchain-based property & rental records and transactions could dramatically increase the speed and transparency of property transactions while reducing the cost of transactions
- **Logistics and Supply Chain Management**[65]: blockchain enables supply chain to detect counterfeit components,

---

[65] Raja Wasim Ahmada et al, Blockchain for Aerospace and Defense: Opportunities and Open Research Challenges, Computers & Industrial Engineering Journal, November 2020

locate the spare parts requiring Maintenance, Repair, and Operations (MRO), and establish the provenance of each part. This is particularly beneficial in industries where transparency and traceability become critical, such as aerospace, military and food industry

- **Enable IoT solutions:** in IoT networks, the exchange of information is critical. Cyber-physical-systems must be able to exchange Information quickly and safely. IOTA has been specifically designed for this purpose.

Conclusion

In this book, the main technologies associated with Industry 4.0 have been described.

These technologies are enabling the implementation of Smart factories. Therefore, knowing each of them represents a fundamental requirement towards the evolution of production systems, in which greater flexibility is guaranteed without compromise with costs, quality and times.

Flexibility represents the new paradigm of smart factories, which can only be achieved through technology, mainly of a digital nature. However, technology represents a necessary but not sufficient condition: as underlined in the introductory chapter, an Intelligent Factory cannot ignore human skills, i.e., the ability to integrate these technologies in a synergistic way with the workforce, the product, the available budget and spaces, just to name a few variables.

I hope the contents of this book have lived up to expectations. If so, I invite you to enter a positive evaluation within the Amazon store!

Thank you,

Nicola Accialini

List of Figures

Figure 1: FDM 3D Printer Extruder 9

Figure 2: How Photopolymerization works 14

Figure 3: Schematic representation of Inkjet Technology 16

Figure 4: How Binder Jetting works 18

Figure 5: How Laser Sintering works 23

Figure 6: Example of 3D printed support for final inspection 37

Figure 7: Microsoft HoloLens Headset 52

Figure 8: Industrial robot cost decline 60

Figure 9: A drone 78

Figure 10: Storage supply and demand from 2006 to 2020 . 87

Figure 11: Explosion of digital data from 2006 to 2020 Source: Patrick Cheesman 88

Figure 12: Levels of Data Science 95

Figure 13: Service models in Cloud Computing 114

Figure 14: Server Racks 124

Figure 15: Vertical and Horizontal integration under Industry 4.0 171

Figure 16: Block scheme of a Discrete Event Simulation process steps 189

Figure 17: Pictorial representation of CAVE technology 202

Figure 18: HTC Vive Headset 206

Figure 19: Global Revenue from Smart Wearables and Hearables 214

Figure 20: concept representation of a Digital Twin 224

Figure 21: How blockchain works 229

Figure 22: Organization of Tangle blocks 233

List of Tables

Table 1: Material Extrusion VS Photopolymerization 15

Table 1: ABB IRB 1400 Yumi Data Sheet 72

Table 2: Comau Aura Data Sheet 73

Table 3: Fanuc CE 35iA Data Sheet 74

Table 4: Kuka LBR IIWA 14 R820 Data Sheet 75

Table 5: Universal Robots UR3 / U5 Data Sheets.................. 76

Table 7: How DES can support different phases in manufacturing.. 187

www.ingramcontent.com/pod-product-compliance
Lightning Source LLC
Chambersburg PA
CBHW071448220526
45472CB00003B/716